Studies in Industrial Chemistry

For CSE and O-level

Bill Harrison and David Wright

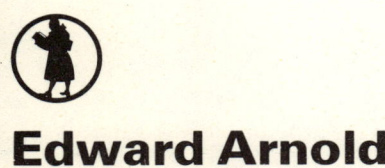

Edward Arnold

First published 1982
by Edward Arnold (Publishers) Ltd
41 Bedford Square
London WC1B 3DQ

British Library Cataloguing in Publication Data

Harrison, W.
  Industrial chemistry.
  1. Chemistry, Technical
  I. Title   II. Wright, D.J.
  660   TP155

ISBN 0 7131 0588 7

Acknowledgements

The Publishers wish to thank the following for their permission to use copyright photographs:

Esso	cover
Documentation Française	p. 12 top
B.P.	p. 12 bottom
B.S.C.	p.p. 34 & 36
BBC TV	p. 42
British Aluminium Co.	p. 43

Typeset by Oxprint Ltd, Oxford
Printed by Thomson Litho Ltd

Contents

Preface

In today's world we depend to a very great extent on the products of the Chemical Industry. It is only right therefore, that we should have some basic understanding of the important chemical manufacturing processes, the raw materials and resources needed to operate them and the social and economic implications that the Chemical Industry has for our lives. Did you know, for example, that over four-fifths of the chemicals we use are produced from petroleum, that several thousand million pounds are lost each year fighting corrosion in the UK alone, or that only ten years ago one chemical complex was polluting our atmosphere by releasing 120 tonnes of sulphur dioxide each day? Incidentally, that has now been reduced to a mere 3 tonnes per day!

There has been a most welcome move in Chemistry Education towards placing a greater emphasis on the applications of chemistry and stressing the wider social implications of the subject. This is also being reflected in the syllabuses and examination papers of several examination boards.

We have chosen seven main topics for study:

The Petroleum Industry
Chemicals from Salt
The Iron and Steel Industry
The Aluminium Industry
Ammonia and Nitric Acid
Sulphuric Acid
Fertilisers

In addition to providing up to date details of the manufacturing processes and the chemistry associated with these, we have tried to present interesting information on each topic. Attention has been drawn to current issues in the Chemical Industry which affect us all, such as the energy crisis and the need for conservation, the particular problems of certain industries, e.g. iron and steel, the importance of fertilisers, the siting of chemical plant, and environmental pollution.

It is important for anyone, regardless of whether or not they are going to work as chemists or engineers, to be able to analyse and think logically. There are therefore a good many examples where information and data are supplied in the form of graphs, pie charts and tables with questions to answer on these at the end of each chapter. You will also be asked to think about and reach decisions on such problems as the siting of a chemical plant or the operation of a process.

The chemist has a responsibility to protect the environment by ensuring that waste products are carefully monitored and controlled and that the process does not seriously interfere with the balance of nature and of the natural beauty of the landscape. He must also ensure the safety of those who work the process or live nearby. A good deal of attention has been given to these aspects and we hope you will enjoy the role playing and discussion suggestions included to highlight these.

In some chapters 'Optional Study' topics have been provided to either give more specific details on a particular aspect or to introduce points of general interest associated with the case study. Your teacher will advise on these optional studies.

Throughout the book the American billion has been used, thus

1 billion = 1 000 000 000.

Note on the questions

The shorter questions are followed by bracketed numbers that indicate the text section, figure or table in which relevant information may be found, thus

(S1.5) refers to Section 1.5
(F1.5) refers to Figure 1.5
(T1.5) refers to Table 1.5.

1 The chemical industry

1.1 The importance of the chemical industry

The modern Chemical Industry undoubtedly makes a most important contribution to our society. The products of the industry are so widespread that they find applications in almost every aspect of our lives (see Figure 1.1).

From its early beginnings as small localised companies producing colour dyes for textiles, tanning leather and making glass, the Chemical Industry has become one of the fastest growing industries and is now the second largest in the United Kingdom.

It is the earth and its atmosphere that provide the vital fuels, raw materials and water supplies upon which the industry depends. Britain's main natural resources are coal, oil and natural gas, limestone, sand, salt, iron ore, fluorspar, gypsum, anhydrite, china clay, barytes and small amounts of the ores of tin, lead and strontium. It was in areas that had ready supplies of these resources that the industry was first established, e.g. the salt based industries which grew up around the Cheshire salt deposits.

Other factors also influenced the introduction of a manufacturing process such as a local market for the products, availability of cheap labour, dock facilities for imported raw materials and a good transport network. Today cheap industrial land

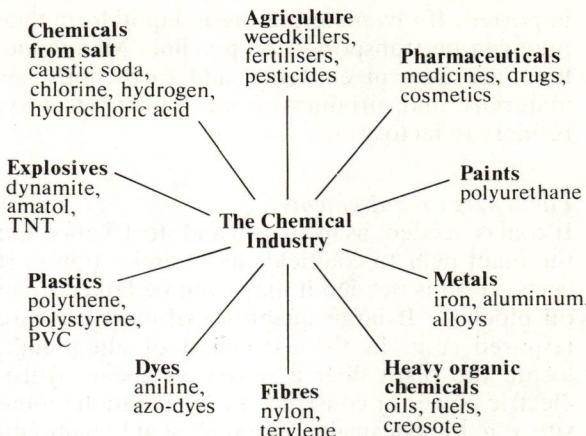

Figure 1.1. The Chemical Industry engages in a great variety of activities which can be broken up into ten main categories.

away from heavily populated areas and government grants to attract companies to development areas are just as likely to influence industrial development. It is worth while considering some of the social and economic factors which must be taken into account when setting up a chemical plant, but first look at Figure 1.2 which shows the main stages in a chemical manufacturing process.

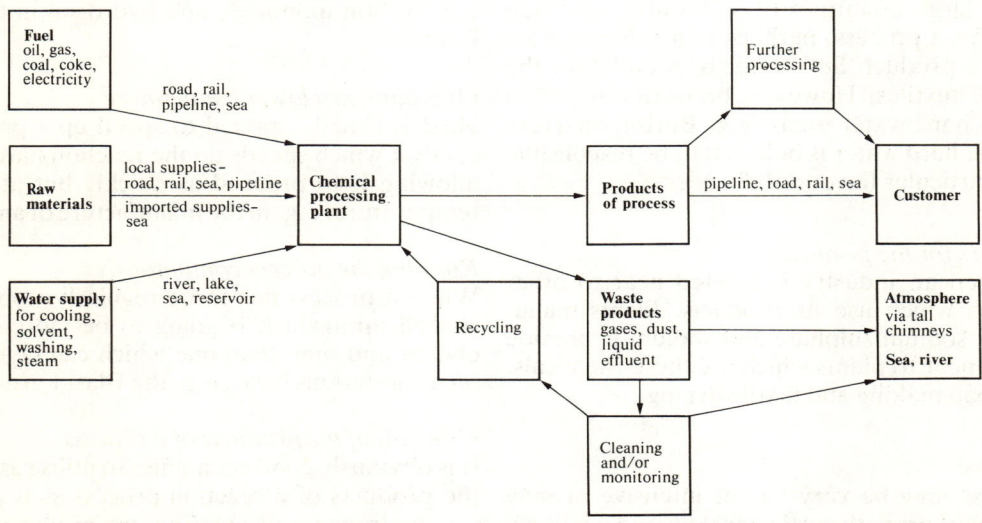

Figure 1.2. Raw materials, fuel and a supply of water are the main requirements for most chemical processes. The products of the process may be further processed before going to the customer. Waste products may be cleaned and recycled or disposed of into the atmosphere, sea or river.

1

1.2 Social and economic factors which influence the location of a chemical plant

Raw materials
It makes good sense to site the plant near to natural deposits or near to a port if materials are imported. If raw materials are in liquid form then they can be transported by pipeline. Many pipelines are now in existence and carry both raw materials and products from port to factory, refinery to factory etc.

Fuel/energy requirements
If coal is needed, as in an iron and steel works, site the plant near to coalfields to minimise transport costs. If oil is needed it may even be linked to an oil pipeline. If large quantities of electricity are required, e.g. in the extraction of aluminium, locate the plant near a source of cheap hydro-electric power or construct a power station on the site, e.g. the Alcan aluminium plant at Lynemouth in Northumberland.

Water supplies
Large amounts of cooling water are used by the Chemical Industry, e.g. in steel making, and this is a very important consideration. Siting near a river, lake or sea is usually necessary. Occasionally it may be necessary to provide new water resources, e.g. the building of a reservoir in Upper Teesdale in the North-east of England to supply the chemical plants of Teesside.

Fairly large quantities of soft water are often needed for a process, perhaps as a solvent or for washing a product. Soft water is essential for the dyeing of textiles. However, breweries are often found in hard water areas, e.g. Burton-on-Trent where the hard water is believed to be responsible for the particular flavour of the beer.

Customers for the products
Very often an industry is located near to other industries which use its products. Plants manufacturing sodium sulphate and sodium hydroxide are sited near to plants which use these chemicals, e.g. in soap making and textile dyeing.

Workforce
A process may be very labour intensive or may require workers with particular skills and it is likely that the plant will be located in an area which can satisfy these requirements. Companies are often attracted to a development area, which is likely to have high unemployment, by government grants or interest-free loans, e.g. the Alcan aluminium plant at Lynemouth in Northumberland.

Transport network
The availability of good road, rail or canal networks can influence the location of chemical plant. The chemical industry of Merseyside uses the Manchester Ship Canal facilities and has also benefited from the construction of the M6 Motorway. The location of the Wilton works on Teesside was probably partly influenced by the fact that it could be linked to the already established Billingham works by a 16 km long chemicals pipeline (as wide as a London Underground tube tunnel).

Potentially dangerous processes
Processes which produce toxic waste or products or use highly flammable or even explosive materials, e.g. nuclear power stations, metal and oil refining plants, should be sited away from heavily populated areas.

1.3 Minimising energy costs – energy conservation

Transfer of energy
With an exothermic reaction the heat given out can be transferred to another part of the process with the aid of heat exchangers. Heat can also be removed from hot waste gases before being expelled. Some waste gases can be used as fuels, e.g. carbon monoxide and hydrogen in the Blast Furnace.

Operating at a lower temperature
Heat is usually applied to speed up a process. A catalyst which speeds up the reaction may be used allowing it to proceed as quickly but at a lower temperature, e.g. in the manufacture of ammonia.

Running the process continuously
When a process has to be regularly stopped and started up again it is going to be more costly in energy and time than one which can be operated on a continuous basis, e.g. the Blast Furnace.

Using all of the products of a process
It is obviously good economics to utilise as many of the products of a reaction process as is possible, e.g. hydrogen and chlorine are produced during the manufacture of sodium hydroxide. These are too valuable to waste and have important uses.

1.4 Environmental considerations

The Chemical Industry contributes to both atmospheric and water pollution which can be very harmful to all forms of life.

The industry expels a good deal of smoke, dust and noxious gases into the air, and waste and toxic chemicals into our rivers, lakes and seas. Everyday necessities such as detergents, indestructible plastics and the motor car contribute to the pollution problem. Many people are only too aware of the disasters due to chemical pollution, such as the mercury poisoning at Minamata in Japan in 1953 and the leakage into the air over Seveso in North Italy of a deadly poison in 1976.

However, although it is vitally important to learn from our mistakes, it must be remembered that used wisely chemicals have given us immense material benefits. With care and effort many of the problems of pollution can be minimised and even eliminated. Nowadays a great deal of effort is being made by the Chemical Industry to monitor and control the levels of harmful wastes released from its plants. These levels are also monitored by government inspectors.

1.5 The role of the Chemical Industry in the world

The Chemical Industry today makes a major contribution to a country's economy and plays a vital role worldwide. The flow of its raw materials and products requires international co-operation, the most modern scientific techniques, skilled management and financial investment. The USA has the largest Chemical Industry in the world, far outstriding its nearest rivals: Japan, West Germany, the UK and France. The combined Chemical Industry of the EEC now almost matches that of the USA.

Those countries with highly developed Chemical Industries supply the less developed nations with the vital chemical products they need such as fertilisers, medicines, fibres, plastics and chemical intermediates. In addition they can provide expertise in chemical plant design and operation.

The modern manager in the Chemical Industry needs to have a variety of skills to be able to liaise with many departments and operations (see Figure 1.3). In addition he/she may need to have the ability to work at an international level.

1.6 Questions on the text sections

1 Make a list of Britain's main natural resources. (S1.1)

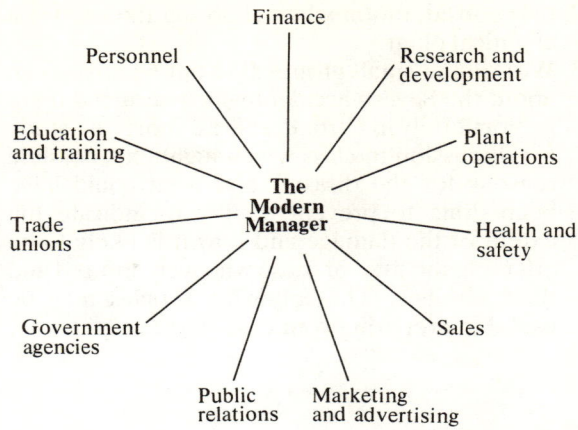

Figure 1.3. The modern manager in today's Chemical Industry needs many skills and talents to cope with the demanding and varied nature of the work.

2 Why did the salt based industries grow up around Cheshire? (S1.1)
3 What probably helped to influence the location of the Wilton Works on Teesside? (S1.2)
4 Why can the use of a catalyst help to reduce the energy costs of a process? (S1.3)
5 Name three everyday causes of pollution. (S1.4)
6 Which country has the largest Chemical Industry in the world? Give reasons why you think this should be. (S1.5)

1.7 Questions on the figures

1 What are the ten main sections of the Chemical Industry? (F1.1)
2 Study Figure 1.2 and then try to show the main stages in a chemical manufacturing process.
3 List the departments or sections of a chemical company with which a chemical engineering manager may have to work. (F1.3)

1.8 Longer questions

1 In order to improve our standard of living it will be necessary to expand the Chemical Industry.

Expansion of the Chemical Industry inevitably leads to pollution and the destruction of the environment.
(a) Are either, or both of these statements necessarily true? Give reasons.
(b) Using these two statements build a case either *for* or *against* the expansion of the Chemical Industry.

3

2 Discuss briefly the factors which must be taken into consideration when choosing the site for a chemical plant.

3 Working in small groups find out what you can about the Seveso accident which occurred there in North Italy in 1976. Prepare a short case study for discussion in class which highlights possible reasons for the disaster and what could have been done to prevent it. Try to indicate the extent of the damage and how it is likely to be affecting the lives of those who were injured and their families. The following articles may be useful in preparing your case studies.

'Seveso – the accident that should never have happened', *New Scientist*, 5 August, 1976.

'After Seveso, What?', *New Scientist*, 12 August, 1976.

'The Graveyard on Milan's Doorstep', *New Scientist*, 19 August, 1976.

'Seveso', *New Scientist* Letters, 2 September, 1976.

'Concealed Data', *New Scientist* Letters, 9 September, 1976.

'Seveso', *The Guardian*, 28 March, 1977.

2 Petroleum

2.1 Introduction

Petroleum (oil and natural gas together) is one of the most valuable and versatile materials found in the earth. Petroleum and its products are essential to the development of industrial nations, providing them with most of their energy requirements for heating, transport and industry. It is used in the manufacture of fertilisers, pesticides and foodstuffs and for making many hundreds of products used in everyday life.

2.2 What is petroleum?

No one knows for sure just how oil and natural gas were formed hundreds of millions of years ago. However, at that time most of the living shell-like creatures were found in shallow seas. When they died they fell to the bottom of the sea and sank in the mud mixing with the dead sea plants and fresh water plants carried down to the sea by the rivers. More mud and sand was deposited, burying these creatures and plants deeper and deeper in the sea bed. Over millions of years they were compressed and this organic matter was converted to oil. The earth's crust also changed, some parts, once under the sea, being pushed up above sea level. This resulted in oil and gas being trapped in porous layers of rock (see Figure 2.1).

Crude oil is a complex mixture of **hydrocarbons** (compounds of hydrogen and carbon only) with small amounts of impurities, chiefly sulphur, with

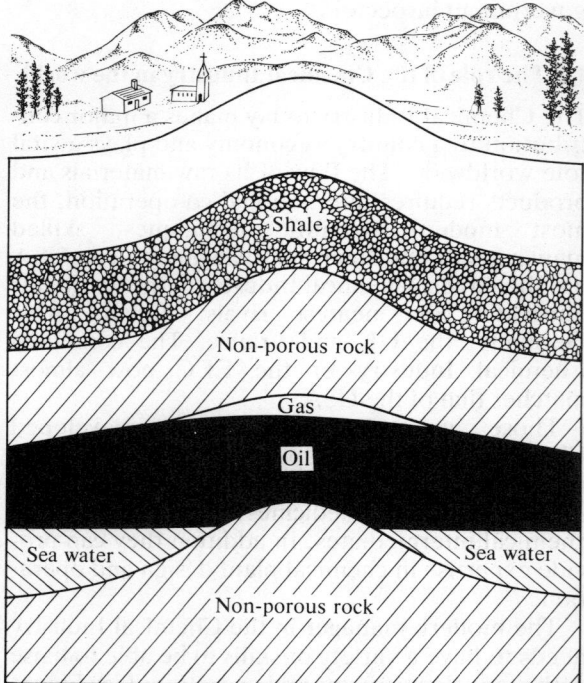

Figure 2.1. Oil and gas is trapped in a porous layer of rock with non-porous rock above and below it.

traces of oxygen, nitrogen, water and some metals. Natural gas is made up of about 90% **methane**, which is the simplest hydrocarbon, having the formula CH_4.

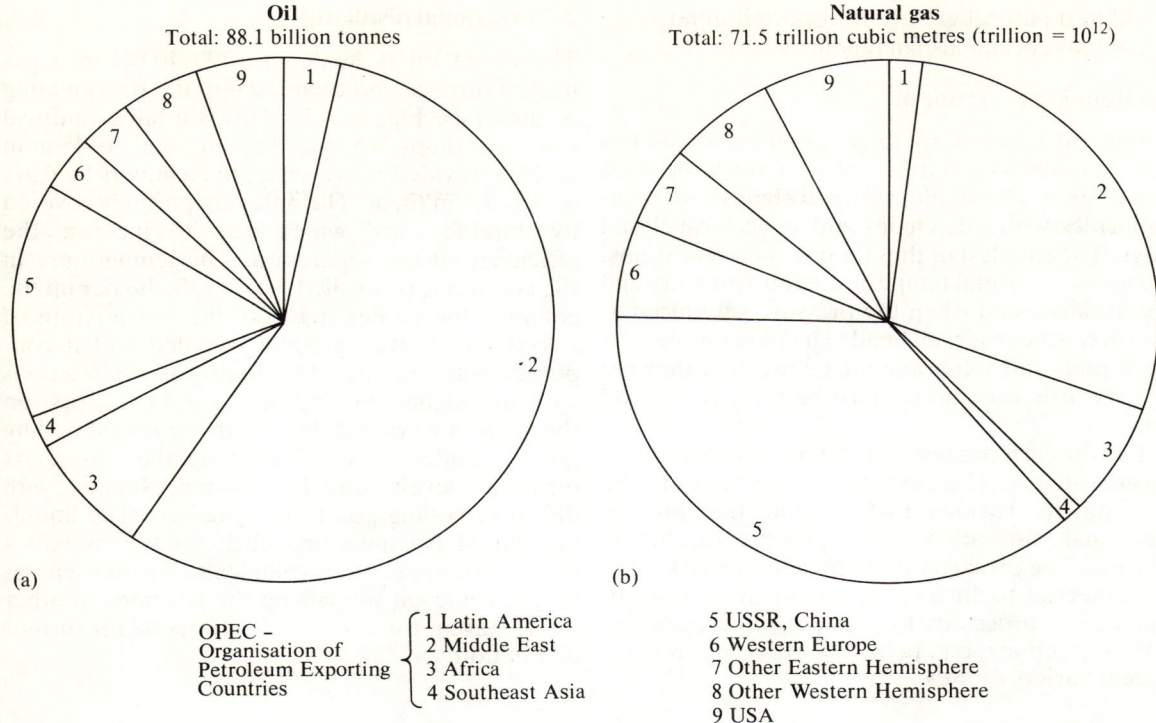

Figure 2.2. Published proven world reserves of (a) oil and (b) natural gas at the end of 1978.

2.3 Where petroleum is found

Petroleum is found in many different parts of the world but a large proportion is concentrated in certain regions (see Figures 2.2 a,b).

Petroleum is found in the UK both inland and offshore (see Figure 2.5).

The total UK (on and offshore) production in 1979 was just over 76 million tonnes whilst the total world production in the same year was over 3 billion tonnes!

2.4 Historical development

As far back as the year 6000 B.C. shallow pits were dug in the ground so that oil which had oozed up from underground could be collected. The pitch from these sticky pools was used to waterproof wooden ships and to bind together bricks for building.

It was not until the eighteenth century that drilling for oil began in the USA as the demand for oil for oil lamps increased.

However, with the invention of the internal combustion engine there was a great demand for petrol, which led to the large expansion of the Industry between 1910 and 1935. The increased demand for petrol and the development of the Petrochemical Industry resulted in further expansion.

Today the Petroleum Industry is concerned with:

Searching and drilling for petroleum

Transporting the petroleum

Refining the crude oil

Oil products and markets

Chemicals from petroleum

2.5 Drilling for and transporting petroleum

Today drilling for oil and natural gas is a very costly and highly technical business, requiring the combined skills of geologists, who study the rocks and their formations, and geophysicists, who are concerned with the mechanical, electrical, magnetic and gravitational properties of the rocks. They use very sophisticated equipment and carry out aerial photography, seismic surveys (i.e. setting off a small explosion underground and recording echoes that rebound from the different layers of rock), and even exploratory drilling on land and sea, in the search for oil.

5

Oil and natural gas are transported in huge sea going tankers and also in pipelines.

2.6 Refining the crude oil

Crude oil consists of large numbers of hydrocarbon molecules of different sizes and structures. Some exist as simple straight chains, some as molecules with side chains and some form closed rings. The smallest of these, i.e. 1–4 carbon atoms, are gases at normal temperatures and pressure and are usually freed when the pressure is reduced as the oil reaches the well head. The larger molecules are liquids and solids and mixed together they are of very little use and so must be refined at an oil refinery.

The modern process of refining is carried out in several stages. The first stage is to separate the oil into its various hydrocarbon fractions by **fractional distillation** – the primary distillation process. The products are then either used directly or subjected to further treatment by a series of conversion processes to produce either more of certain fractions, e.g. petrol, or other products for a great variety of uses.

2.7 Fractional distillation

The crude oil is heated to 300–400°C in a gas heated furnace and then fed into the fractionating column (see Figure 2.3). This is a tall cylindrical tower (perhaps 3–8 m in diameter and 30–45 m in height), divided into a series of chambers by trays with holes in them. The holes are partially covered by 'bubble caps' which help to increase the efficiency of the separation. The temperature in the column is controlled, so that the higher up the column, the cooler it is. As the hot mixture of vapours rises it is gradually cooled and it condenses into liquids. The hydrocarbon fractions with the highest boiling points condense first on the bottom trays and those with the lowest boiling points condense last, higher up the tower. At different levels up the column, liquids with different boiling points are collected. The liquids are called **fractions** and each fraction is still a mixture of many other chemicals. Further separation is achieved by passing the fractions to other fractionating towers or 'side strippers' for further distillation.

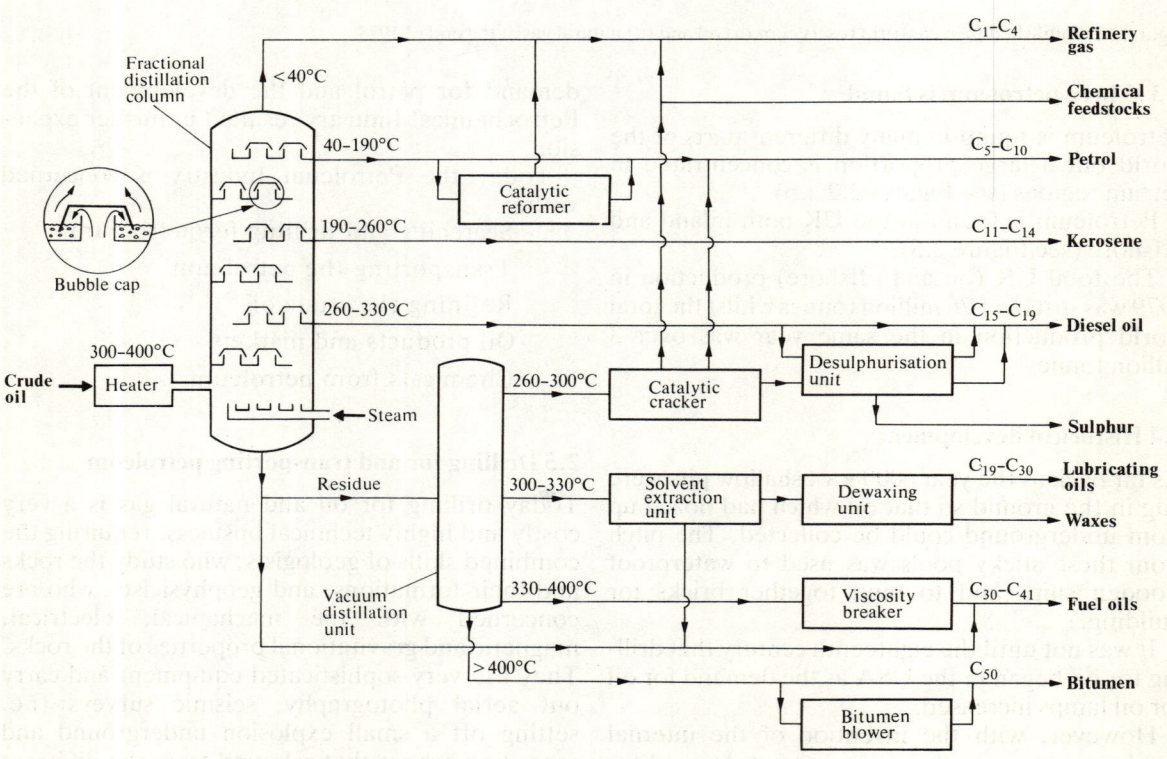

Figure 2.3. The processing of crude oil into the many products we use involves several different processes all of which are carried out in a modern oil refinery.

2.8 Vacuum distillation

Low in the tower, the heaviest hydrocarbon fractions flow to the bottom as a viscous mass or residue. This residue is fed into a vacuum distillation unit where it is distilled under reduced pressure, since the high temperature required to vaporise its components would decompose them. The fractions obtained are used for:

feedstock for catalytic cracking (discussed later).
the manufacture of lubricating oils and waxes (candles, polishes).
producing rubbery bitumen (roofing felt, mastic) and hard bitumen (roads).

2.9 Catalytic cracking (Cat-cracking)

The petrol content of any given crude oil is not very great (see Figure 2.4) and in any case, the

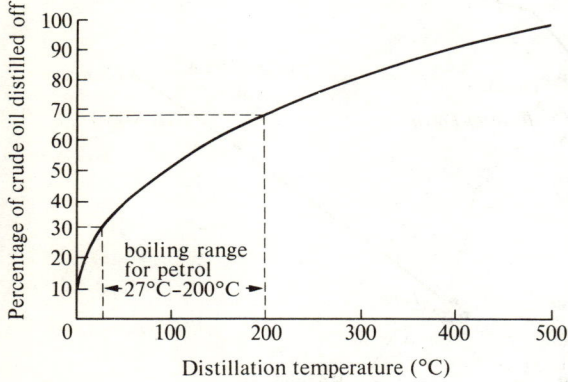

Figure 2.4. Petrol boils off from crude oil within the small temperature range shown on the graph.

petrol obtained directly from the distillation column is not suitable for modern engines. With the rapid growth of the motor car a way had to be found to break down the heavier hydrocarbon fractions into the lighter petrol fractions. This was achieved by a process known as **cracking** which has made it possible to produce from a barrel of crude oil more than twice as much petrol as is possible from distillation, that is also of a higher quality.

In addition the cracking process also provides molecules suitable for use as chemical feedstocks (i.e. materials from which chemicals are manufactured).

Consider a fairly heavy hydrocarbon molecule containing a chain of 10 carbon atoms which when subjected to a 'cracking' process breaks up to give 2 smaller (and different) types of hydrocarbon molecules:

decane

heptane
(suitable for
petrol)

propene
(used as a
chemical
feedstock)

Hydrocarbons such as heptane have better **anti-knock** properties than those produced directly by distillation, i.e. they give smoother engine performance without 'knocking' or 'pinking' and can be used in engines with higher compression ratios, which give more power for their size.

Catalytic cracking requires a temperature of about 500°C and an aluminium oxide/silicon catalyst. The catalyst provides a surface which aids and speeds up the cracking process. Cracked petrol is blended with other lower grade petrols to improve their quality.

2.10 Catalytic re-forming (plat-forming)

This is another method used to improve the quality of petrols, by causing the molecules of straight chain hydrocarbons in low grade petrols to break up and **re-form** as 'branched chain' hydrocarbons, which produce higher grade petrols, e.g.

octane

platinum
catalyst

high pressure
and temperature

iso-octane

7

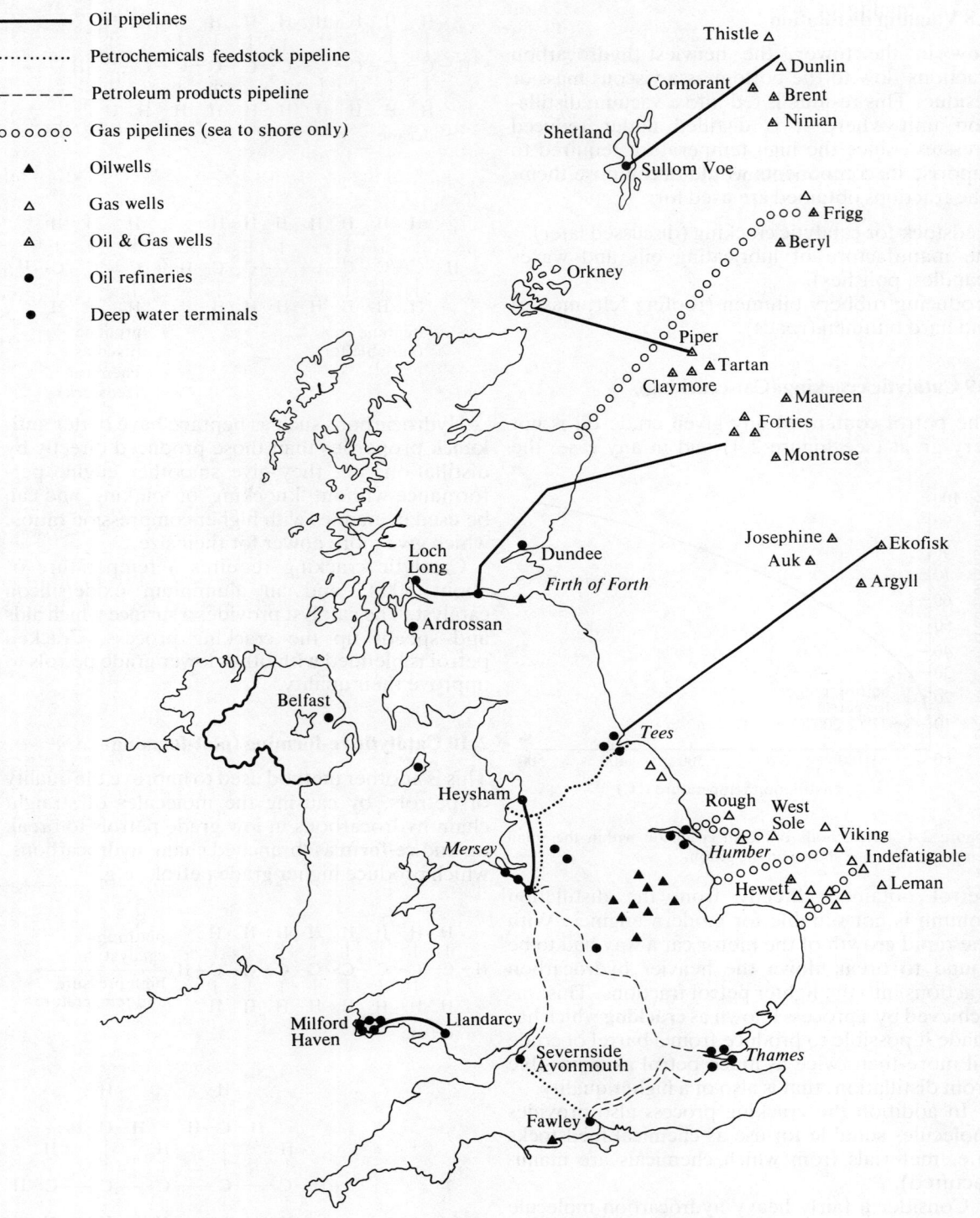

Figure 2.5. The location of oil and gas wells (both inshore and offshore in the UK), refineries and the petroleum and petroleum products distribution network.

8

The quality of a petrol is given by its **Octane Number**. Iso-octane is given the number 100 and all other petrols are compared to this. Two star petrol has an Octane Number of 92 and is only 92% as efficient as pure iso-octane. The higher the Octane Number the better the petrol's 'anti-knock' properties.

Re-forming to produce molecules with ring structures is also carried out to produce both better quality petrols and chemical feedstock.

2.11 Other conversion processes

Other conversion processes involve the linking together of small molecules (e.g. refinery gases) to form larger molecules (e.g. petrol). The gases are subjected to high pressures and temperatures in the presence of a catalyst which forces them to unite or **polymerise** and form liquids called **polymers** (see Figure 2.9). Polymers are essential components of high octane motor and aviation fuels.

2.12 Location of oil refineries

A modern refinery is a huge complex perhaps covering an area up to 2000 acres and refining up to 20 million tonnes of crude oil per year. Figure 2.5 shows the location of the main refineries in the UK.

The Refinery Industry grew up in the 1930s around the main estuaries in the UK since the oil tankers of that time, being of about 12 000 tonnes capacity, could use the major ports. However, as the demand for oil grew and super-tankers were built weighing over 200 000 tonnes, special tanker

terminals had to be built, e.g. Loch Long and Milford Haven. In more recent years Europe's largest oil terminal was constructed at Sullom Voe in the Shetlands at a cost of over £900 million and will soon handle well over half the nation's oil requirements, which will be channelled to the Voe in two 100-mile pipelines.

The oil has to be transported from the terminals to the refineries and this is nowadays often done by underground pipeline (see Figure 2.5). Nearly all of the refineries are located on the coastline or near large estuaries. They require large areas of cheap land not too near heavily populated areas. These installations are potentially highly dangerous and there is always the risk of fire, explosion and pollution of the atmosphere and coastline. However, they cannot be sited too far from large areas of population because the oil and its products are needed for factories and homes. There has been a great deal of opposition to the location of certain refineries, such as at Milford Haven, not only for the dangers mentioned above, but also because of the effect such a large installation has on the natural beauty of such a coastline.

The UK is nowadays one of the most important refining areas in the the world refining almost 150 million tonnes of crude oil per year.

2.13 The uses of petroleum and its products

The two most important uses of petroleum are as a source of fuel to provide energy and as a source of chemicals. It is therefore worth looking a little more closely at these particular uses (see Figure 2.6).

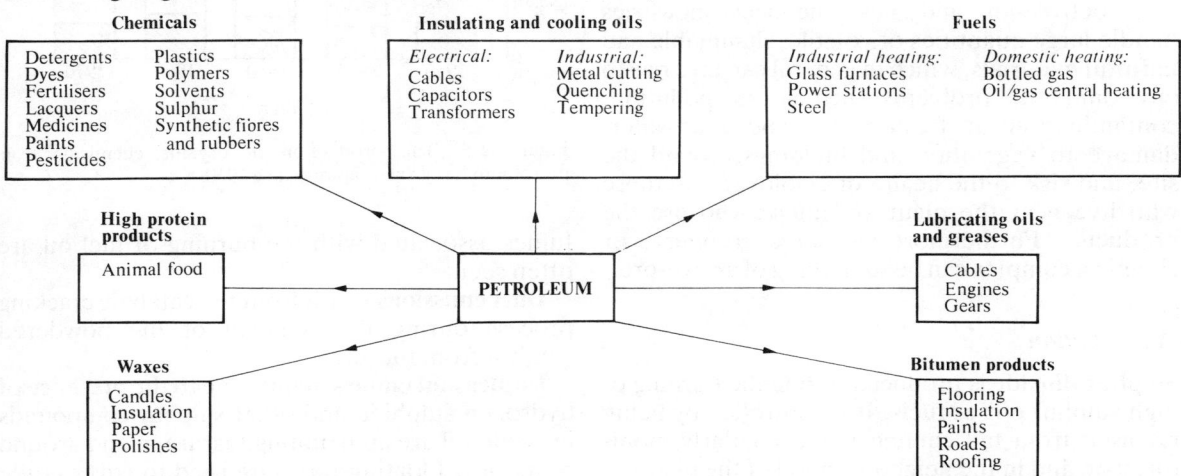

Figure 2.6. Petroleum and its products have many important uses. In many cases they form the starting materials for the preparation of other products.

2.14 Fuels from petroleum

A look at Figure 2.7 shows that about 50% of the world's energy demand is met by oil and 20% by natural gas. Even though the world's consumption of energy has increased by about four times over the last 30 years, most of this increase has been met by petroleum fuels. It is important to realise that we will continue to rely on petroleum as our major energy source for the foreseeable future.

2.15 Chemicals from petroleum

The importance of the Petrochemical Industry in our society cannot be overstated. It really began in the 1930s in the USA and Figure 2.8 shows its rapid growth between 1950 and 1970. Nowadays, over 90% of all organic chemicals are produced from petroleum, the other main source being coal.

The main chemicals manufactured from petroleum are shown in Figure 2.6. Petroleum chemicals are produced chiefly from the hydrocarbon gases ethene (C_2H_4) and propene (C_3H_6), which are obtained by cracking the oil fractions described earlier. Other chemical feedstocks are natural gas, waste refinery gases and aromatic hydrocarbons (e.g. benzene and toluene) which are also extracted from the oil fractions. From these simple substances hundreds of different chemicals (mainly organic) are manufactured (see Figure 2.9).

Sulphur is also produced as a by-product from crude oil and natural gas (see Chapter 4).

2.16 Environmental considerations

The petroleum and petrochemical industries handle large quantities of volatile, flammable and harmful materials, which in general can give rise to environmental problems such as air pollution, contamination of fresh water and sea water, damage to vegetation and buildings around the site, and risk to the health of employees, to those who live near the plant and those who use the products. Furthermore a large refinery or chemical complex can be something of an eyesore.

Air pollution

Sulphur dioxide is produced during the burning of high sulphur oils as fuels. It is controlled by being released from tall chimneys and regularly monitored on and in the neighbourhood of the plant.

Black smoke or **soot** is produced occasionally, due to an upset or unusual flaring of gases. White

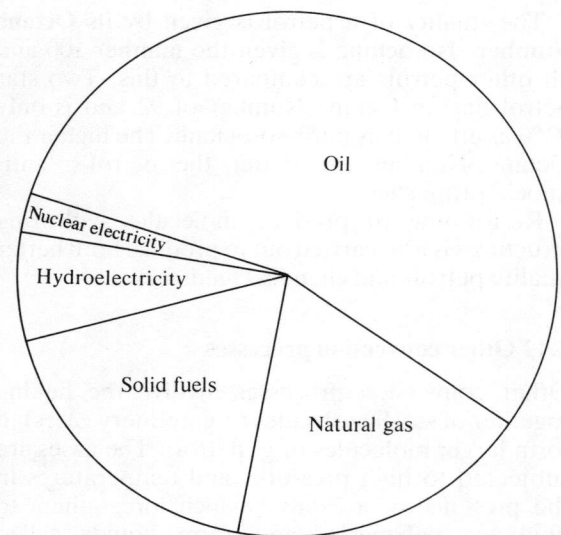

Total energy demand equivalent to 4203 million tonnes of oil

Figure 2.7. World commercial energy demand in 1978. It is easy to see that most of the demand is met by petroleum.

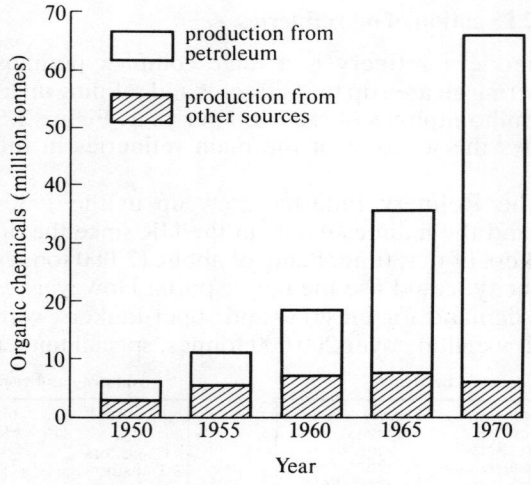

Figure 2.8. The production of organic chemicals from petroleum has grown rapidly since 1950.

fumes associated with the burning of fuel oil are often seen.

Dust emissions result from the catalytic cracking process during the removal of the powdered catalyst from the gases.

Unpleasant odours mainly due to the presence of hydrogen sulphide and other sulphur compounds in crude oil are an irritating feature in and around refineries. Floating roofs are used to cover crude oil storage tanks to help reduce evaporation and odour emissions.

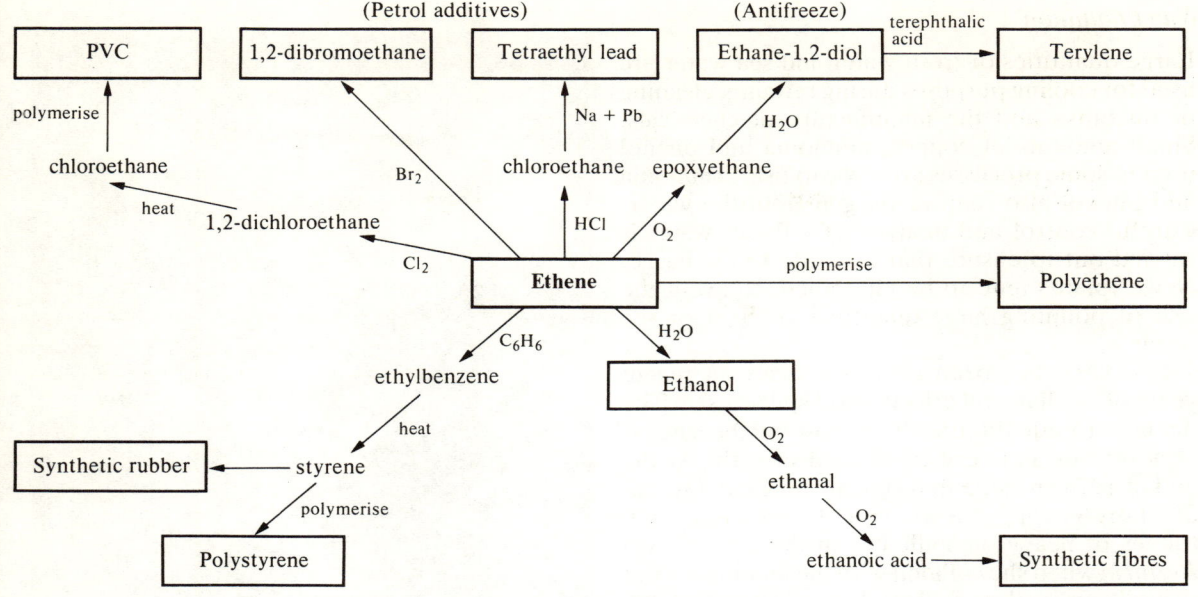

Figure 2.9. Ethene is the starting material from which many petrochemicals are made. Note how many of these chemicals are very important in our everyday lives. Products in boxes are important chemicals.

Lead pollution is becoming quite a serious problem, especially in large cities where the emissons of exhaust gases from motor vehicles are great (see Table 2.1). Exhaust gases, besides releasing carbon monoxide, also emit lead, which is added to petrol in the form of tetraethyl lead to reduce 'knocking' or 'pinking'. Lead is very toxic, particularly to small children living near motorways or in big cities. Pressure is being applied to governments and oil companies to stop the addition of lead to petrol and rather to produce higher octane petrol.

Pesticides (e.g. D.D.T.) and **herbicides** (e.g. 245-T) manufactured from petroleum chemicals are toxic and can damage the health of both animals and humans. 245-T used for crop spraying in the USA and Australia has been linked with miscarriages and birth deformities.

Table 2.1 Average daily intake of lead by a 'normal' person in the USA (from WALKER, C. 1971: *Environmetal Pollution by chemicals*. Hutchinson Educational).

Substance ingested	Daily intake	Lead concentration in substance	Lead ingested per day (mg)	Fraction absorbed	Lead absorbed per day (mg)
Food	2 kg	0.17 p.p.m.	330	0.05	17
Water	1 kg	0.01 p.p.m.	10	0.1	1
Town air	20 m³	1.3 mg/m³	26	0.4	10.4
Country air	20 m³	0.05 mg/m³	1	0.4	0.4
Tobacco smoke	30 cigarettes	0.8 mg per cigarette	24	0.4	9.6

Water pollution

Large quantities of fresh water and sea water are used for cooling purposes during refining, cleaning of oil tanks and the manufacture of chemicals. Small amounts of copper, ammonia and phenol used in some processes are toxic to fish. Ammonia and phenol also remove oxygen from the water. Careful control and analysis of effluent water is carried out to ensure that it is safe to discharge. Newer plants tend to be air cooled, to avoid the risk of polluting large quantities of river or sea water.

Everyone has been only too aware in recent years of the harmful effects of oil spills at sea. The damage to aquatic life, birds and our beaches is very serious and very costly to deal with. At the end of 1978 an estimated 6000 sea birds and scores of shore-grazing sheep were killed after 1160 tonnes of heavy oil spilled from the tanker *Esso Bernicia* when she collided with the loading jetty at the oil terminal at Sullom Voe. Sections of the Shetland coastline are still polluted in spite of a £3½ million cleaning operation.

Modern terminals have very sophisticated control equipment, including radar and aerial sur-

The *Amoco-Cadiz* wrecked off Portsall, Finistère in 1978 spilling out 230 000 tonnes of crude oil into the sea

veillance. The risk of collisions at sea can be reduced by fitting tankers with radar and advanced navigational aids. Techniques for dealing with oil spills by using oil booms and dispersants are improving, but it is still a frightening thought to consider a 500 000 tonne super-tanker spilling its load near to our shores.

2.17 The energy crisis

At the end of 1979 the proven world reserves of petroleum were estimated at about 100 billion tonnes but if we continue to consume it at the present rate it is thought likely to run out in the next 30 years. A look at Table 2.2 shows that there was a three-fold increase in the consumption of petroleum between 1940 and 1960 and between 1960 and 1978. If this trend continues then we are likely to run out even earlier!

Fortunately new reserves of oil have only just been discovered in Venezuela, which are estimated to almost double the world's proven reserves. Added to this are finds in the North Sea, Alaska and almost daily announcements of promising oil prospects in the Gulf of Mexico, off the Cameroons, off China or even in the approaches to the English Channel, and we may conclude that there is plenty of oil in the earth. The energy crisis

Pollution clean-up after the *Esso Bernicia* incident at Sullom Voe terminal in 1978.

Table 2.2 World petroleum consumption.													
Year	1940	1950	1960	1968	1969	1970	1971	1972	1973	1974	1975	1976	1977	1978
Consumption (billion tonnes)	0.3	0.5	1.0	1.9	2.1	2.3	2.4	2.6	2.8	2.7	2.7	2.9	3.0	3.1

is not then one of a shortage of energy reserves but rather one of a shortage of time. We cannot find petroleum and get it out quick enough to match the rate at which we are using it up. In order to keep pace we would need to discover every year as much oil as has been found in the North Sea in the last 10 years! So there still is a very real energy crisis. We must all conserve energy and especially limit the use of petroleum, since reference to Figure 2.7 shows us that for the foreseeable future it is by far and away our major energy requirement. It is also our major chemical feedstock. Even though the British Government plans to make large investments in nuclear energy production, it will take a good many years before that source can supply all our needs. Furthermore, there are many who would prefer not to develop this source of power and see it as being potentially highly dangerous.

2.18 The future

For the present and the future we must put great efforts into energy conservation and the development of alternative energy sources such as hydrogen from the electrolysis of water; solar, wind and wave power; heat from the interior of the earth; oil from coal (we have large supplies). Serious consideration must even be given to American Professor Melvin Calvin's idea of the 'Petrol Tree'. He has discovered that the sap from the gopher tree which grows in arid or desert areas, contains about $\frac{1}{3}$ hydrocarbons. It is like crude oil and can be turned into petrol and plastics. By cultivating large areas of desert with these trees, he estimates that petrol could be produced for about 3p per gallon.

Also, recent research at Manchester University has led to the discovery of a process for converting a city's rubbish into good quality crude oil. 1000 tonnes of oil can be produced from 2000 tonnes of rubbish and it is estimated that the process could provide about 3% of Britain's energy requirements.

In an attempt to achieve fuel economy in cars, British Leyland have produced the ECV-2 (the

The Power of Oil

O.P.E.C. (The Organisation of Petroleum Exporting Countries, mainly made up of Middle East Arab states and N. African countries) control the majority of oil production and prices. The power of this organisation to affect our lives was painfully demonstrated, when they decided to use oil as an economic and political weapon after the renewed outbreak of war between Israel and the Arab states in 1973. An oil embargo was introduced and within three months of this action the world demand for oil exceeded the supply. As well as controlling the supply, OPEC also quadrupled the price of oil. Since western Europe and Japan have become almost totally dependent on cheap OPEC oil for sustaining their economies, this action had a disastrous effect. Firstly it greatly increased inflation because they depend so much on oil and oil products and secondly it was a major factor in starting up the world recession, both of which cause unemployment.

The 'political' power of oil must also be recognised. Oil producing countries, by controlling the production of oil can effectively hold to ransom those countries which depend on oil for their very survival.

Furthermore, much of the world's oil supplies are concentrated in the Middle East, an area of great political unease, and therefore strategically it is of vital importance since any power controlling that area would also control most of the world's vital energy supplies.

Figure 2.10. We should all be fully aware of the power of oil to influence our lives.

Experimental Conservation Vehicle No. 2) which has an aluminium and plastic body and is capable of doing 100 miles per gallon! The company is also developing LPG cars which run on the much cheaper liquefied petroleum gas from the North Sea. By the late 1980s B.L. hope to be producing cars which are 50% more fuel efficient than today's models.

2.19 Questions on the text sections

1 Describe briefly with the aid of a diagram how petroleum was originally formed. (S2.2)
2 What do you understand by the term 'hydrocarbon'? (S2.2)
3 Which hydrocarbon is present in the highest proportion in natural gas? Give its name and formula. (S2.2)
4 What was the main reason for the Oil Industry 'taking off' in the eighteenth century? (S2.4)
5 What was the reason for the large expansion of the Oil Industry during the years 1910–35? (S2.4)

6 State three methods by which petroleum deposits can be located. (S2.5)

7 How is crude oil transported from (a) oil well to port? (b) port to port? (c) port to refinery? (S2.5)

8 Complete the missing words represented by the letters A to I.

A fractional distillation column separates the crude oil into fractions each with a different (A). . . . range. Each fraction is made up of hydrocarbons with only certain numbers of (B). . . . atoms in the chain. The result is that the lighter, (C). . . . boiling point liquids reach the (D). . . . of the column and the heavier, (E). . . . boiling point liquids distil off at the (F). . . . of the column. Petrol, which is the most volatile product, comes off at the (G). . . . of the column as a vapour, which is then (H). . . . to a liquid. The residue from the bottom of the column is fed into a (I). . . . distillation unit for further separation. (S2.7)

9 Name three fractions, other than petrol, which are obtained during distillation. (S2.7)

10 Why was the process of 'cracking' of the heavier oil fractions introduced? (S2.9)

11 Why are some petrols catalytically reformed? (S2.10)

12 What is the advantage in using high octane petrol? (S2.10)

13 (a) Petrochemicals are chiefly produced from which two hydrocarbon gases?
   (b) What are their formulas?
   (c) State some other feedstocks from which petrochemicals are produced. (S2.15)

2.20 Questions on the figures and tables

1 At the end of 1978 total world reserves of oil were about 88 billion tonnes. How much of that was contributed by Western Europe? (F2.2a)

2 Assuming that recent discoveries of oil in Venezuela (Latin America) are confirmed as being equal to the total proven world reserves in 1978, redraw the pie diagram (Figure 2.2a) to show the new situation.

3 What is the temperature range over which petrol boils? What percentage of the total crude oil fractions would therefore be petrol? (F2.4)

4 If the total world energy demand in 1978 was equivalent to 4203 million tonnes of oil, work out the number of tonnes contributed by each of: solid fuels, hydro- and nuclear electricity. (F2.7)

5 Work out for each year shown the proportion of organic chemicals produced from petroleum. (F2.8)

6 (a) From which of the substances listed in Table 2.1 is most lead ingested?
   (b) From which of the substances listed is most lead actually absorbed?
   (c) How many times is the concentration of lead in town air greater than in country air? What do you think is the reason for this?
   (d) Who would absorb the most lead – a smoker living in the country or a non-smoker living in the town?

7 Study the data in Table 2.2 and work out the increase in consumption of petroleum during the periods 1940–60 and 1960–78. Is this a three-fold increase in each case?

2.21 Longer questions

1 Give a brief account of the catalytic cracking and reforming processes, giving example equations.

2 Look at the map in Figure 2.5 and answer the following questions.
   (a) Why are the refineries located around the coastline?
   (b) Can you suggest a reason why the inland oil fields are found grouped around the Midlands?
   (c) What are the advantages of underground pipelines for transporting oil, gas and chemicals?
   (d) *Either* Working on your own, write a short account giving the reasons for and against the siting of a large oil refinery at Milford Haven.

   Or Form two small discussion groups: one representing the residents and those with business interests in tourism in Exmouth, Devon, and the other representing the interests of a large petroleum company which proposes to build a large, new refinery there to serve the needs of the south-west of the country. The local residents group will try to give strong arguments against building there, whilst the other group will argue the case for that location. Plan your arguments as separate groups and then come together to debate.

3 List the main sources of pollution of the environment arising from the operations of the Petroleum Industry. How does the Industry attempt to control (a) waste gases and dust? (b) odour emissions? (c) lead emissions?

4 What do you think would be the full effects of a serious oil spill fairly close to our coastline?

5 Even though recent oil discoveries in Venzuela and other parts of the world could double reserves why do you feel it is necessary to conserve energy?

6 Discuss briefly the alternative sources of energy of which you are aware.

7 Use the data in Table 2.2 and plot a graph of world consumption of petroleum between 1940 and 1978. Project your graph as near as you can to the existing slope up to the year 2000.

(a) Work out how much oil was consumed between 1940 and 1960 and between 1960 and 1980 and how much is likely to be consumed between 1980 and 2000.

(b) If world reserves were confirmed to be even 200 billion tonnes, and no more was discovered after 1980, how much would be left in the year 2000?

3 Ammonia and nitric acid

3.1 Introduction

Ammonia is one of the most important chemicals being manufactured today. As the world population gets larger, more and more crops will be needed to feed it and as land is being used up for industry and housing, the available farmland will have to be used with the greatest efficiency. This means that fertilisers are needed and, as natural fertilisers alone cannot supply this need, more artificial fertilisers must be made. Artificial fertilisers are made from ammonia.

Between 1962 and 1985 the amount of money spent by developing countries on fertilisers to feed their population will have to rise from £1400 million to £16 500 million. Figure 3.1 shows the relationship between the size of the world population and the world production of ammonia.

3.2 Historical development

Nitrogen compounds are essential for life because the element nitrogen is found in all proteins: the 'building blocks' of most animal and some plant material. Only a few plants (e.g. clover, peas and beans) can take in nitrogen directly from the air. The rest need to take in nitrates from the soil, while animals can only absorb the element by eating plants or other animals. The circulation of nitrogen between the soil and living material is called the **nitrogen cycle**.

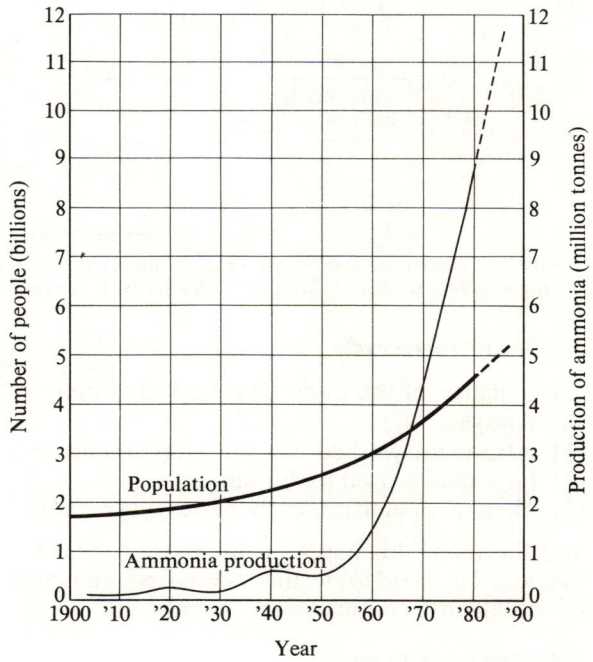

Figure 3.1. Graph showing the population explosion this century which has made it essential to increase the amount of ammonia produced. Broken lines indicate predicted figures.

15

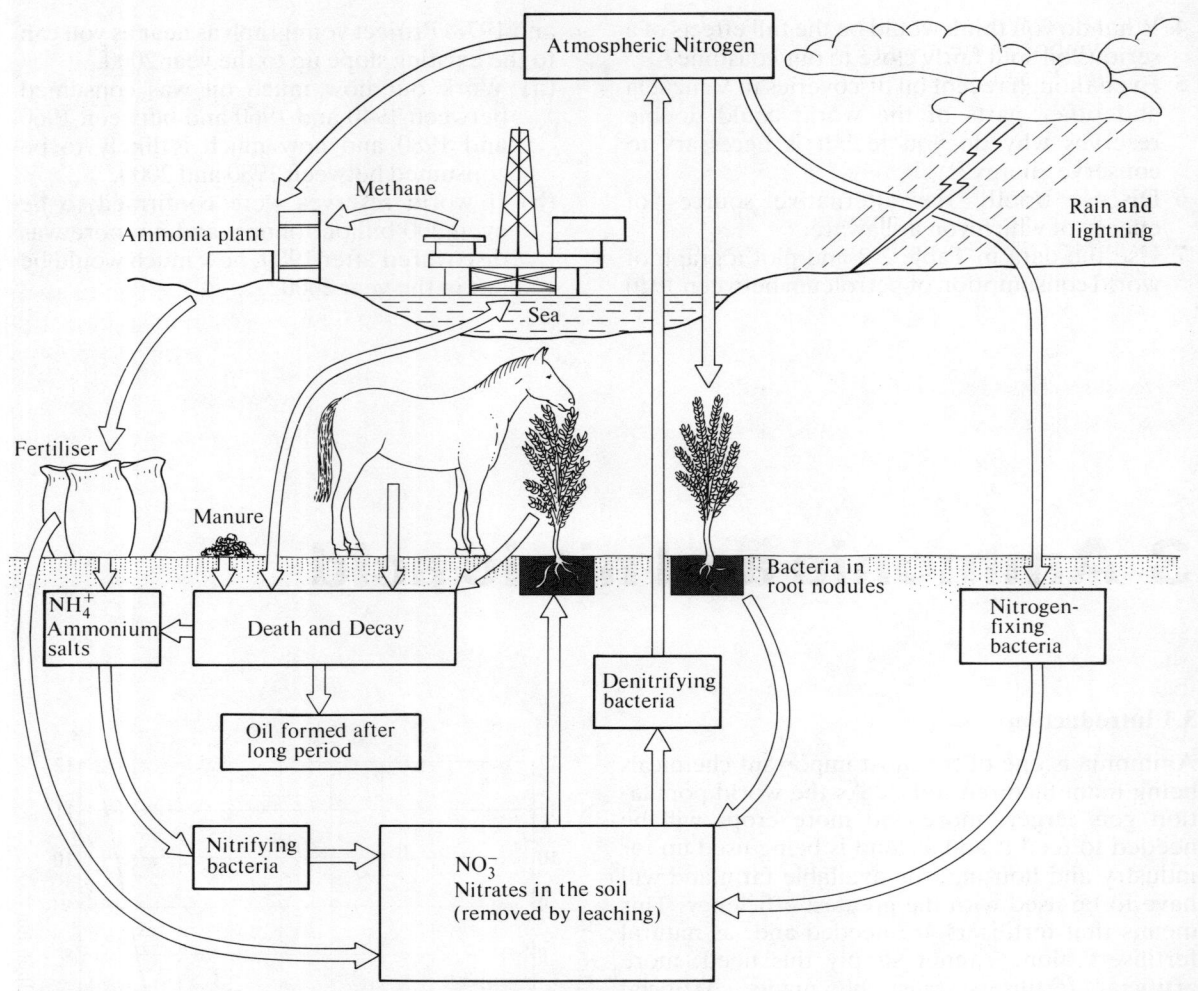

Figure 3.2. The Nitrogen cycle, showing the importance of ammonia. Nitrogen is removed from the cycle when nitrates are washed away and plants are destroyed. It can only effectively be replaced by fertilisers.

3.3 The nitrogen cycle

The balance of the cycle (Figure 3.2) is upset in three ways:

(1) nitrates are washed away by rivers and streams
(2) farm land is used for buildings
(3) plants are eaten rather than decaying naturally

and the more this happens, the more fertiliser is needed. The need for fertilisers was first realised in the nineteenth century.

3.4 The Haber Process

The great breakthrough was made by Fritz Haber in Germany. In the years leading up to World War I, Germany was afraid that imports of Chile saltpetre would be blockaded by the British navy and needed an alternative supply of fertilisers.

The main problem in converting atmospheric nitrogen to ammonia is to break the strong bond between the atoms in nitrogen gas, N_2. This was finally done using a high pressure and a **catalyst**. A catalyst speeds up a chemical reaction but does not change itself. The catalyst used was osmium (a dense metal) and the hydrogen came from coal and steam. The initial **yield** (the percentage of nitrogen actually turned into ammonia) was 10–14%.

3.5 The chemistry of the Haber Process

The reaction between nitrogen and hydrogen is **reversible** and **exothermic** (heat is given out).

$$N_2(g) \; + \; 3H_2(g) \; \rightleftharpoons \; 2NH_3(g) + \text{heat}$$
$$\text{nitrogen} \quad \text{hydrogen} \qquad\qquad \text{ammonia}$$

The symbol ⇌ means that the reaction is reversible, i.e. it can take place in either direction. When nitrogen and hydrogen react, ammonia is formed. This then **dissociates** (breaks down) to form nitrogen and hydrogen again. After a time, the reaction vessel contains definite amounts of nitrogen, hydrogen and ammonia. The reactions still take place, but the overall amount of each stays the same. This is called **chemical equilibrium**.

3.6 Finding the best conditions

The amount of ammonia produced depends on the temperature and the pressure inside the reaction vessel. The reason for this is explained by **Le Chatelier's Principle**.

The highest yield of ammonia is obtained with a **high pressure** and a **low temperature**. Very high pressures are expensive to produce and need a strong reaction vessel. Low temperatures make the reaction very slow. The best conditions have been found to be

temperature 380–450°C
catalyst iron
pressure 250 atmospheres
promoter to prevent the catalyst being poisoned

3.7 The modern manufacturing process

The modern process has been developed by I.C.I. at Billingham. The raw materials are:

nitrogen from the air
methane from North Sea gas
water

Altogether, the ammonia plants in Britain make a total of about 7000 tonnes per day of liquid ammonia.

The plants are sited close to sources of **energy** (coal, gas or oil), **water** (river or sea), **transport** (sea, road and rail) and **labour**. In Britain ammonia plants are sited at Billingham on the River Tees, at Immingham on the River Humber, at Ince Marsh on the River Mersey, and at Avonmouth on the River Severn.

Figure 3.3 shows how the fuel used can affect the cost per tonne of the ammonia produced. At first the price fell due to more efficient processes, but inflation has caused a steady rise since the mid-sixties.

Figure 3.4 shows the route of the ammonia making process in a modern plant.

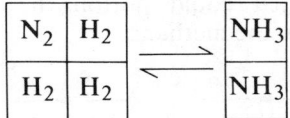
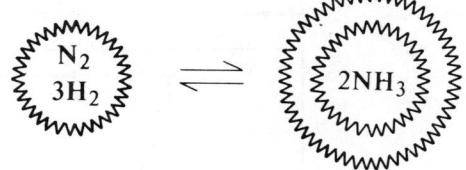

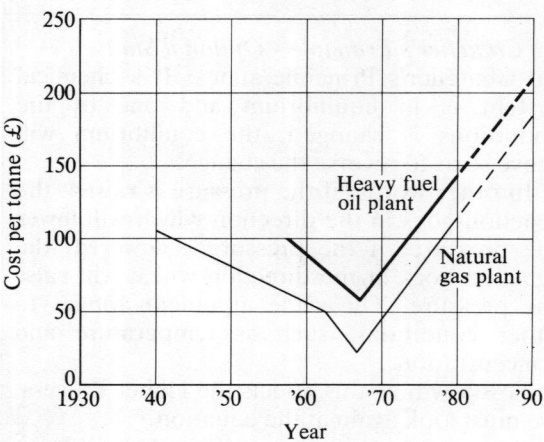

Figure 3.3. Variations in the cost per tonne of ammonia this century. Note that different fuels can make a large difference to the cost. Broken lines indicate predicted figures.

3.8 Producing hydrogen

See Figure 3.4.

Stage 1: Desulphurisation
Sulphur, which could poison the catalyst, is removed from the methane.

Stage 2: Primary steam reformer
The methane is reacted with steam. The methane is oxidised to carbon monoxide and the steam is reduced to hydrogen.

$$CH_4(g) \; + \; H_2O(g) \rightleftharpoons CO(g) \; + \; 3H_2(g)$$
methane   water   carbon   hydrogen
          (steam)   monoxide

temperature 750°C **pressure** 30 atmospheres
catalyst nickel

Stage 3: Secondary steam reformer
Air (78% nitrogen, 21% oxygen) is blown in. The oxygen burns the hydrogen to water, raising the temperature to 1100°C. Nearly all the methane is removed and the nitrogen and hydrogen are left in the ratio 1 volume nitrogen to 3 volumes hydrogen. Carbon monoxide and some carbon dioxide are also present, with the water vapour.

Stages 4 and 5: Water gas shift reactors
The carbon monoxide is oxidised to carbon dioxide and the water is reduced to hydrogen in two reactors.

$$CO(g) \; + \; H_2O(g) \rightleftharpoons CO_2(g) \; + \; H_2(g)$$
carbon   water   carbon   hydrogen
monoxide   dioxide

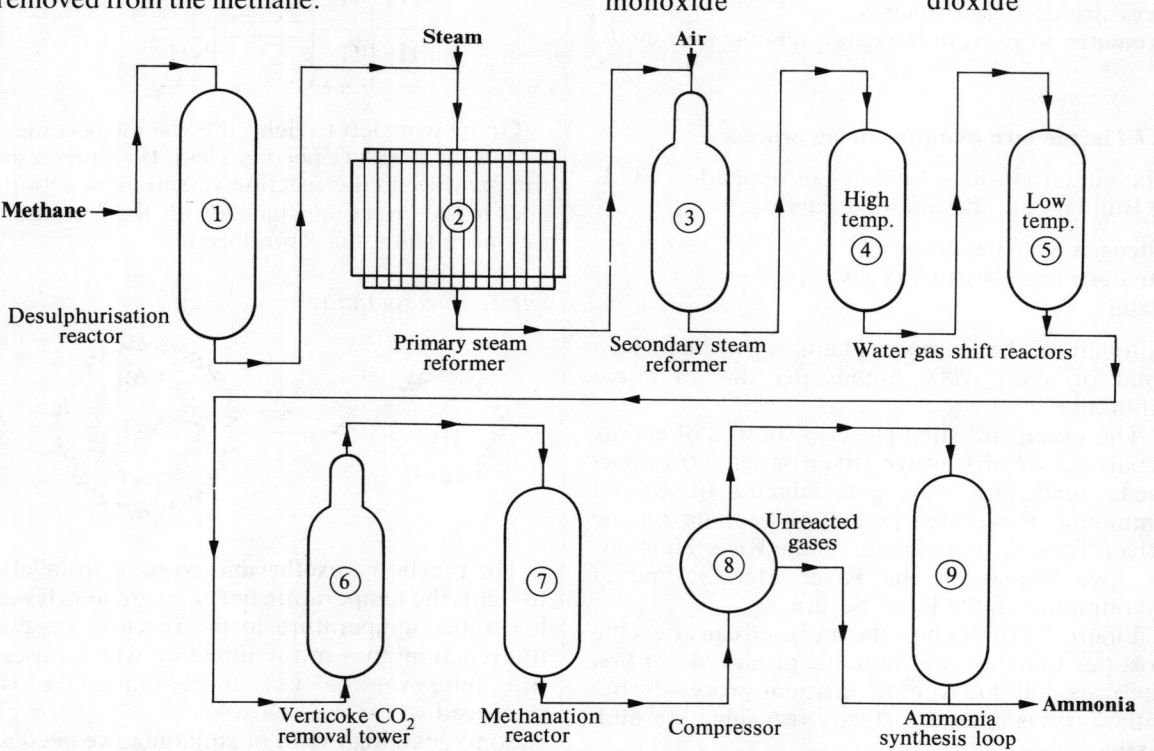

Figure 3.4. The route in a modern ammonia plant. Only the important steps are shown. Numbers refer to the stages mentioned in the text.

18

reactor 1 (high temperature) 400°C, iron(III) oxide catalyst
reactor 2 (low temperature) 220°C, copper catalyst

Stage 6: Carbon dioxide removal
The carbon dioxide is absorbed in hot potassium carbonate.

Stage 7: Methanation
Any remaining carbon monoxide is converted to methane.

Stage 8: Compression
The **pressure** is raised from 30 atmospheres to 200 atmospheres and the **temperature** is lowered to 380–450°C (the compression raises the temperature).

3.9 Synthesis of ammonia

Stage 9: Ammonia synthesis loop
(Synthesis – building up from the elements.) The mixture of 3 volumes of hydrogen to 1 volume of nitrogen is reacted together. Not all of it forms ammonia at first (see Figure 3.4) but the unreacted gases are recycled until they are mostly combined.

$$N_2(g) + 3H_2(g) \rightleftharpoons 2NH_3(g)$$
nitrogen   hydrogen   ammonia

temperature 380–450°C **pressure** 250 atmospheres **catalyst** iron + promoters

The reaction is exothermic and produces enough heat to keep the temperature high enough. Ammonia is removed by **condensation** (water cooling) or dissolving in water, before the unreacted nitrogen and hydrogen are returned to the loop.

The conditions of temperature and pressure in the reaction vessels can make a large difference to the amount of ammonia produced, and thus affect the cost of the process (see Table 3.1).

3.11 Uses of ammonia See Table 3.2.

Table 3.1 The final percentage yield of ammonia from the Haber Process under different conditions of temperature and pressure.

Temperature (°C)	Pressure (atmospheres)		
	150	250	350
350	46.2%	57.5%	65.2%
450	22.3%	32.9%	39.3%
550	9.9%	15.6%	20.8%

3.10 The future

With the present population growth (see Figure 3.1) another 500 ammonia plants will be needed in the next 25 years. Large plants are not economical and the high cost of transport means that plants need to be built near where the ammonia will be used.

New sources of energy may be needed, such as solar energy. Certain crops can also be grown and burnt to fuel the Haber Process. These produce enough fertiliser to compensate for the land lost by growing crops for fuel instead of food!

Although some progress has been made in the study of nitrogen fixing bacteria (bacteria which turn nitrogen from the air directly into nitrates) and incorporating these into cereal crops, artificial fertilisers will still be needed.

Research is constantly taking place to try and improve the efficiency of the present process, reducing the amount of energy required and the overall size of the plant. Improvements can be made with:

(1) more efficient catalysts to save energy.
(2) better design to increase efficiency.
(3) new equipment to reduce size.

Table 3.2 The uses of ammonia (NH_3). Simple reactions are shown together with the products, some of which are further converted.

Reaction	Product	Uses
$NH_3 + H_2O \rightleftharpoons NH_4OH$	Ammonium hydroxide	Cleaner
$2NH_3 + CO_2 \rightleftharpoons (NH_2)_2CO + H_2O$	Urea	Fertiliser, plastics
$2NH_3 + H_2SO_4 \rightleftharpoons (NH_4)_2SO_4$	Ammonium sulphate	Fertiliser
NH_3 + organic chemicals	Nylon, rayon	Plastics
Cooled NH_3	Liquid ammonia	Refrigerant
$NH_3 + HCl \rightleftharpoons NH_4Cl$	Ammonium chloride	Dry cells, flux
Oxidised NH_3 (see section 3.12)	Nitric acid ↓	Explosives
$NH_3 + HNO_3 \rightleftharpoons NH_4NO_3$	Ammonium nitrate ↓	Fertiliser
	Nitro chalk	Fertiliser

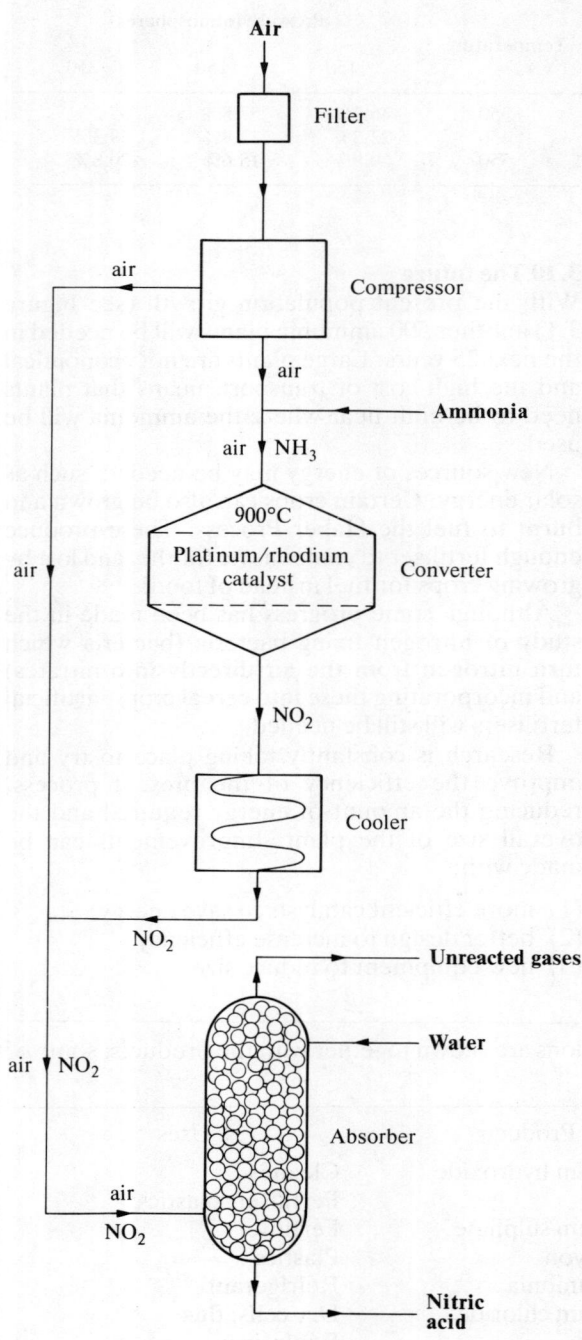

Air

Filter

Compressor

air

air

Ammonia

air NH$_3$

900°C

Platinum/rhodium catalyst

air

Converter

NO$_2$

Cooler

NO$_2$

Unreacted gases

Water

air NO$_2$

Absorber

air

NO$_2$

Nitric acid

Figure 3.5. The manufacture of nitric acid. A simplified flow diagram showing the important stages.

3.12 The manufacture of nitric acid

Figure 3.5 shows a simplified flow diagram of the manufacture of nitric acid. After any dust is filtered out, the air is compressed to several times atmospheric pressure, before ammonia from the Haber Process is added. The high temperature in the converter starts the reaction, which is then exothermic enough to continue on its own.

$$4NH_3(g) + 7O_2(g) \longrightarrow 4NO_2(g) + 6H_2O(g)$$
ammonia   oxygen   nitrogen(IV)   water
oxide

The gases are cooled, mixed with more air, then passed up the absorber against a flow of hot water. Glass balls in the tower give a surface to react on.

$$4NO_2(g) + O_2(g) + 2H_2O(l) \longrightarrow 4HNO_3(aq)$$
nitrogen(IV)   oxygen   water   nitric acid
oxide

The result is 65% nitric acid, 35% water. The acid can be made more concentrated by distillation.

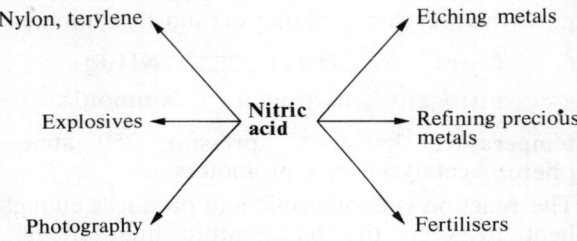

Figure 3.6. The uses of nitric acid. Chapter 5 explains its importance within the fertiliser industry.

3.13 Questions on the text sections

1 Explain why ammonia is important to the world's population. (S3.1)
2 How can the balance of the nitrogen cycle be upset? (S3.3)
3 Why did Germany need to find a new source of fertiliser in the early twentieth century? (S3.4)
4 What is meant by a reversible reaction? (S3.5)
5 What is meant by chemical equilibrium? (S3.5)
6 What ideal conditions are needed for a high yield of ammonia in the Haber Process and why can the ideal conditions not be used? (S3.6)
7 Which three raw materials are used in the modern Haber Process? (S3.7)

8 Write down the word and symbol equation for the reaction between methane and steam. What are the reaction conditions and catalyst? (S3.8)

9 Write down the word and symbol equation for the synthesis of ammonia from nitrogen and hydrogen. What are the reaction conditions and catalyst? (S3.9)

10 How is ammonia removed from the reaction vessel? (S3.9)

3.14 Questions on the figures and tables

1 What was the world population in 1960? (F3.1)

2 How much ammonia was produced in 1970? (F3.1)

3 What is the world population likely to be in 1990? (F3.1)

4 In which ways do nitrates get into the soil? (F3.2)

5 How are nitrates taken out of the soil? (F3.2)

6 Why is ammonia production necessary to the nitrogen cycle? (F3.2)

7 Explain how nitrogen is taken into the human body and how it gets back into the soil. (F3.2)

8 What was the cost per tonne of ammonia from a natural gas plant in 1970? (F3.3)

9 What percentage of the nitrogen/hydrogen mixture would be converted to ammonia at 450°C and 350 atmospheres? (T3.1)

10 Make a list of the products of ammonia and what they are used for. (T3.2)

3.15 Longer questions

1 Copy Figure 3.2 – The nitrogen cycle and ammonia.
Explain:
(a) why nitrogen is important to life.
(b) why it is important that the nitrogen cycle is a cycle.
(c) why fertiliser needs to be put into the cycle.

2 Read the optional study – Le Chatelier's Principle. Explain why an industrial process needs the correct temperature and pressure.

3 Explain why ammonia plants are built where they are, taking into account sources of raw materials, markets for the finished product and sources of labour. Why is it necessary to build plants close to where the product will be needed?

4 Copy Figure 3.4 – Route in a modern plant. Explain in detail what happens at each stage of the process.

5 (a) Use figure 3.5 to explain how nitric acid is manufactured in industry.
(b) Use Le Chatelier's Principle to explain the temperature and pressure used in the oxidation of ammonia.

6 Ammonia is toxic (poisonous) and hydrogen is explosive. Explain the arguments for and against building ammonia plants in heavily populated areas.

4 Sulphuric acid

4.1 Introduction

Although chlorine is probably now a better indicator, it used to be recognised that the quantity of sulphuric acid consumed by a country was a good indication of its prosperity. This can be easily understood if we look at the industries which require sulphuric acid at some stage of the manu-facturing process. Figure 4.1 shows the amount of sulphuric acid consumed in the United Kingdom compared with the rest of the world and Figure 4.2 shows the uses to which it is put.

4.2 Historical development

Sulphuric acid was first produced in A.D. 900. The

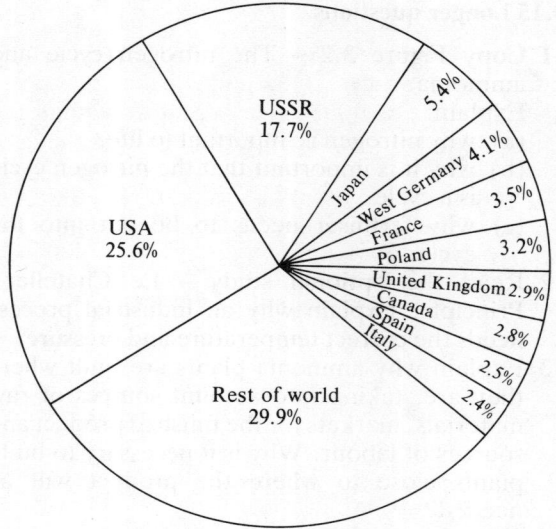

Figure 4.1. Sulphuric acid production in the world. The amount of sulphuric acid produced gives a good indication of a country's prosperity.

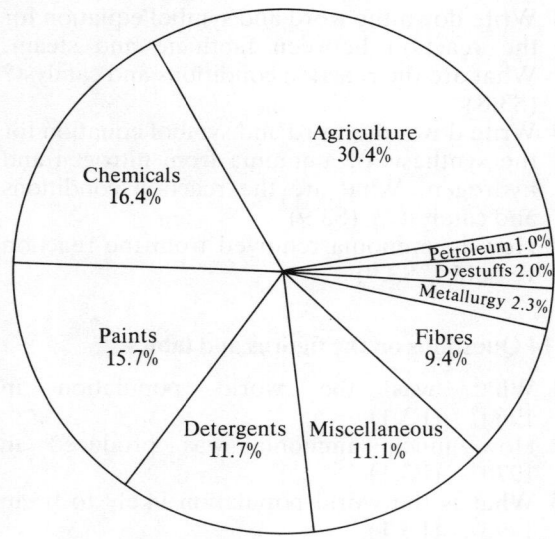

Figure 4.2. Uses of sulphuric acid. Note the importance of the acid in the production of fertilisers.

action of weathering on iron pyrites (FeS_2) produces hydrated iron(II) sulphate ($FeSO_4.7H_2O$). If this is distilled, sulphuric acid is produced. The main uses of sulphuric acid were in cleaning metals and in neutralising the alkalis left after bleaching cloth.

The vast majority of modern sulphuric acid is now produced by the Contact Process.

4.3 The Contact Process

The original Contact Process was invented by Peregrine Phillips in 1931. Phillips passed a mixture of sulphur dioxide and air through a heated tube containing finely divided platinum, producing sulphur trioxide. This was absorbed in water to produce sulphuric acid.

Commercial use of the process was slow because of a lack of understanding of the processes involved, the high cost and the fact that the concentrated acid was not needed. In spite of this, the first plant was opened by Squire and Messel in London in 1875.

By 1901, Le Chatelier's Principle was understood and the dyestuff industry required the concentrated acid. The modern Contact Process is identical to the original one, except that:

(1) a different catalyst is used.
(2) the sulphur trioxide is absorbed in sulphuric acid, not water.

4.4 Sources of sulphur dioxide

95% of the UK's sulphuric acid is made from one of three sources of sulphur dioxide:

(1) pure sulphur from the USA, Mexico or Poland by the Frasch Process.
(2) hydrogen sulphide from France or Canada, found as an impurity in natural gas (15%).
(3) desulphurisation of crude oil (see Figure 2.3).

North Sea gas contains very little sulphur, so it cannot be used.

The other 5% of sulphuric acid comes from the roasting of metal sulphides found as ores.

Anhydrite (calcium sulphate), which used to be mined at Billingham, Cleveland, and spent oxide (hydrogen sulphide from coal gas manufacture) are no longer used for economic reasons.

4.5 The Frasch Process

Figure 4.3 shows the pump used in the Frasch Process. A hole 30 cm in diameter is bored down to a depth of 160 m, then lined with an iron pipe down which the pump is passed. Down the outer tube is passed **superheated water** at a pressure of 10 atm, which allows it to reach a temperature of 170°C. It escapes through holes in the outer collar and the sulphur, which melts at 115°C, forms a liquid. **Compressed air** at 15 atm is passed down the middle tube and forces the sulphur up the second tube. The sulphur is kept liquid by the heat of the water in the outer tube until it reaches the surface, where it is pumped into large vats to solidify. The sulphur is 99.9% pure.

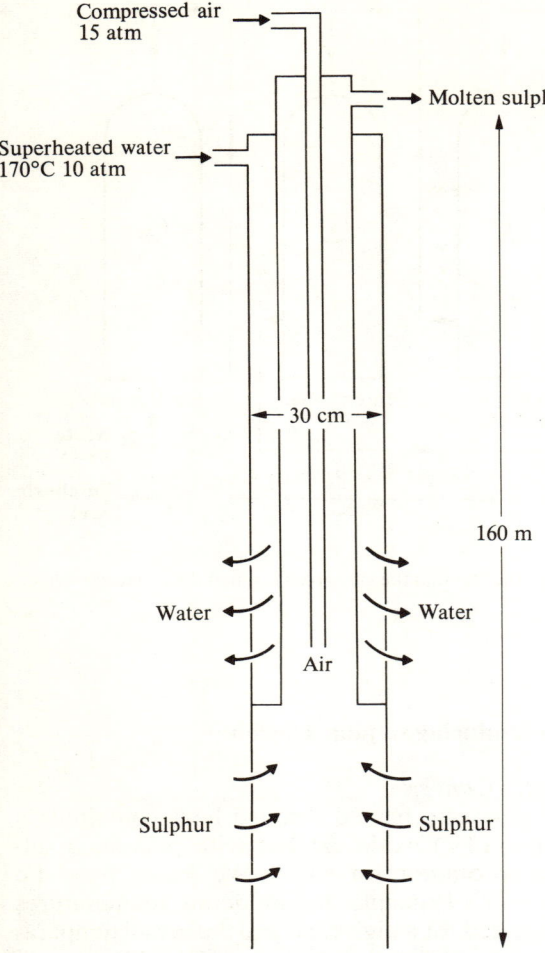

Figure 4.3. The Frasch pump for extracting sulphur. The super-heated water melts the sulphur, which is then blown back to the surface by compressed air.

4.6 Other sources of sulphur

Sulphur from natural gas
Natural gas (methane) can contain about 15% hydrogen sulphide together with some carbon dioxide. If this is passed at 70 atm pressure over an amine solution, the alkaline solution absorbs the acidic gases (carbon dioxide and hydrogen sulphide). If the solution is heated at atmospheric pressure, the gases are released. The hydrogen sulphide is **oxidised** by a controlled amount of air over a bauxite catalyst, producing sulphur which is condensed to a 99.9% pure solid.

$$2H_2S(g) + O_2(g) \longrightarrow 2H_2O(g) + 2S(s)$$
hydrogen   oxygen   water   sulphur
sulphide

Sulphur from petroleum
The petroleum is mixed with hydrogen, then heated. The mixture of gases is passed over a catalyst and the hydrogen combines with sulphur to form hydrogen sulphide. Cooling liquefies the oil and hydrogen sulphide is removed.

To obtain sulphur, the gas is treated as explained in *Sulphur from natural gas*.

4.7 The chemistry of the Contact Process

The sulphur is converted into sulphuric acid in three stages.

(1) The sulphur is burnt in air to form sulphur dioxide.

$$S(l) + O_2(g) \longrightarrow SO_2(g) + \text{heat}$$
sulphur   oxygen   sulphur
                     dioxide

The reaction is **non-reversible** and **exothermic**.

(2) The sulphur dioxide is converted into sulphur trioxide.

$$2SO_2(g) + O_2(g) \rightleftharpoons 2SO_3(g) + \text{heat}$$
sulphur   oxygen   sulphur
dioxide   trioxide

The reaction is **reversible** and **exothermic**. As in the Haber Process, we can use Le Chatelier's Principle to find the best conditions (see Chapter 3 – Ammonia and nitric acid). Once more a high yield can be obtained from a high pressure and a low temperature. In fact the pressures used are only just above that of the atmosphere, because a yield of above 99.5% is possible with a suitable **catalyst**.

(3) The sulphur trioxide is absorbed in water to form sulphuric acid.

$$SO_3(g) + H_2O(l) \rightleftharpoons H_2SO_4(aq) + \text{heat}$$
sulphur   water   sulphuric
trioxide   acid

In actual fact, pure water is not used to absorb the sulphur trioxide. This is because the gas would dissolve in the water vapour above the water, forming a fine mist of sulphuric acid. This would obviously be very dangerous. Instead, the gas is absorbed in a mixture of 98% sulphuric acid and 2% water, which does not produce the mist. The product is **oleum** (pure sulphuric acid), which is not acidic as it contains no ions.

23

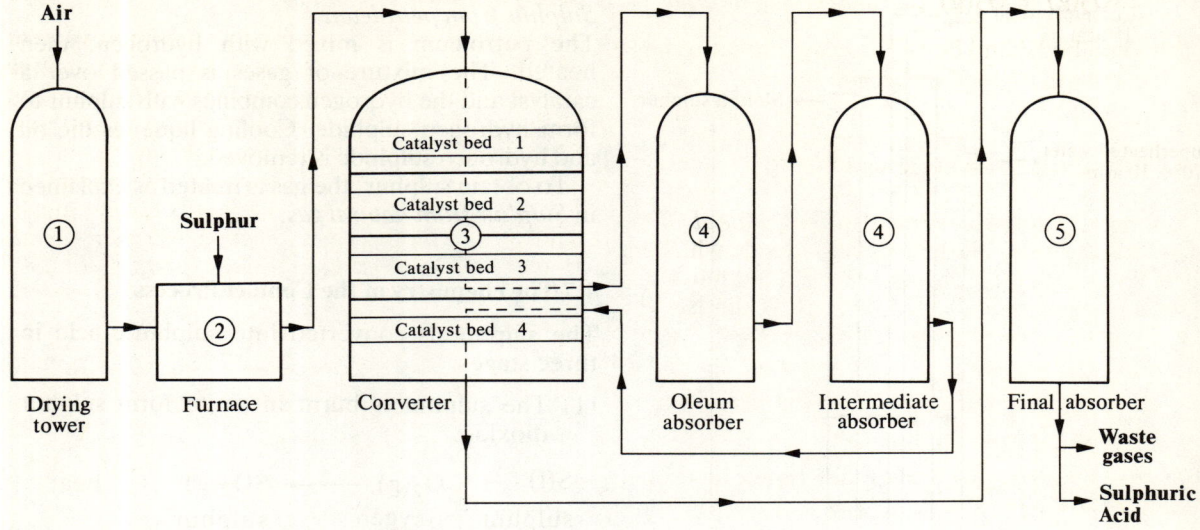

Figure 4.4. The double absorption contact process. The name comes from the fact that the gases pass through the converter twice. Numbers refer to the stages mentioned in the text.

4.8 The modern manufacturing process

Modern plants use a double absorption system, which cuts down the pollution from unconverted sulphur dioxide from 0.18% to 0.05%. To understand how it works we must look at a diagram of the plant (Figure 4.4).

4.9 Producing sulphur dioxide

Stage 1: Drying tower
The air is dried with concentrated sulphuric acid (a **drying agent**) from the final absorber. This prevents an acid mist being formed.

Stage 2: Furnace
Liquid sulphur is burned in air at 1000°C

$$S(l) + O_2(g) \longrightarrow SO_2(g) + heat$$

The energy produced is turned into steam and reused. The mixture of gases which goes into the converter contains:

  10% sulphur dioxide
 0.5% sulphur trioxide
10.5% oxygen
  79% nitrogen

4.10 Producing sulphur trioxide

Stage 3: Converter
The gases are passed through beds containing a vanadium(V) oxide catalyst with potassium sulphate promoter on silica. We know from Le Chatelier's Principle that moderate temperatures are needed for a high yield and the actual temperatures are fixed by the catalyst, which will not work below 400°C and breaks down above 620°C. To keep to these temperatures the gases are cooled after passing through each bed. The temperatures are shown in Table 4.1, together with the percentage of sulphur dioxide which has been converted to sulphur trioxide.

Table 4.1 Temperatures in the converter. Note the high final yield.

Catalyst bed	Temperature of gas before reaction (°C)	Temperature of gas after reaction (°C)	Percentage of SO₂ converted to SO₃
1	435	600	66%
2	445	518	85%
3	445	475	93%
4	420	442	99.5%

$$2SO_2(g) + O_2(g) \rightleftharpoons 2SO_3(g) + heat$$

Between bed 3 and bed 4 the gases are passed into an intermediate absorption system.

4.11 Producing sulphuric acid

Stage 4: Oleum absorber and intermediate absorber
In these towers, all the sulphur trioxide which has been made so far is absorbed in 98% sulphuric acid to prevent acid mist being formed. Water is added to keep this concentration constant.

$$SO_3(g) + H_2O(l) \rightleftharpoons H_2SO_4(aq) + heat$$

Any unconverted sulphur trioxide is passed back to the converter, where it is converted on catalyst bed 4.

Stage 5: Final absorber
Any remaining sulphur trioxide is absorbed in 98% sulphuric acid. Each stage is made as efficient as possible. The final product is used for drying the original air and for dissolving the sulphur trioxide. Water used for cooling is converted to steam as a source of energy, while heat exchangers take unwanted heat to where it is needed.

4.12 The future

The main developments involve the improvement of efficiency. A very high conversion percentage has been achieved and pollution cut to a minimum, although this involves 'wasting' energy as anti-mist filters make the gases flow more slowly.

Higher pressures (at present uneconomical) could improve the yield, although not by much. The large volume of unreactive nitrogen means that the pressure drop in the converter is not 33% but about 8%.

Theory gives:
$$2SO_2 + O_2 \rightleftharpoons 2SO_3$$
$$3 \text{ vol.} \qquad 2 \text{ vol.}$$

$$\frac{\text{drop in}}{\text{volume}} = \frac{3 \text{ vol.} - 2 \text{ vol.}}{3 \text{ vol.}} \times 100\% = 33\%$$

In the industrial process:
$$2SO_2 + \underbrace{2O_2 + 8N_2}_{\text{air}} \rightleftharpoons 2SO_3 + O_2 + 8N_2$$
$$12 \text{ vol.} \qquad\qquad\qquad 11 \text{ vol.}$$

$$\frac{\text{drop in}}{\text{volume}} = \frac{12 \text{ vol.} - 11 \text{ vol.}}{12 \text{ vol.}} \times 100\% = 8.3\%$$

4.13 Uses of sulphuric acid

See Table 4.2 below

Table 4.2 The uses of sulphuric acid (H_2SO_4). Simplified reactions are shown, together with the final products.

Reaction	Product	Uses
$H_2SO_4 + Ca_3(PO_4)_2 \rightarrow Ca(H_2PO_4)_2 + CaSO_4$	Calcium superphosphate	Fertiliser
$H_2SO_4 + 2NH_4OH \rightarrow (NH_4)_2SO_4 + H_2O$	Ammonium sulphate	Fertiliser
$H_2SO_4 +$ wood pulp	Rayon	Fibres
	Cellophane	Film
$H_2SO_4 +$ metal oxides	Metal sulphates	Chemicals
$H_2SO_4 + C_{15}H_{30} + NaOH \rightarrow C_{15}H_{31}SO_4Na$	Detergent	Cleaning
$H_2SO_4 +$ barium and titanium compounds	Pigments	Paints

4.14 Questions on the text sections

1 How does the amount of sulphuric acid made by a country show how prosperous it is? (S4.1)
2 Explain some of the problems involved with the Contact Process before A.D. 1900. (S4.3)
3 Where does the sulphur needed for the Contact Process come from? (S4.4)

4 Look at the Frasch Process for extracting sulphur.
   (a) Why does the water need to be at 170°C?
   (b) Why is the superheated water at high pressure?
   (c) Why is molten sulphur brought up the middle tube?   (S4.5)
5 How is sulphur obtained from natural gas?   (S4.6)
6 Explain what is meant in section 4.7 by:
   (a) a catalyst.   (S3.4)
   (b) a reversible reaction.   (S3.5)
   (c) an exothermic reaction.   (S3.5)
7 Write down the equations for:
   (a) the burning of sulphur in air.
   (b) the conversion of sulphur dioxide to sulphur trioxide.
   (c) the conversion of sulphur trioxide to sulphuric acid.   (S4.7)
8 Which gases pass from the furnace to the converter?   (S4.9)
9 Why are temperatures in the converter kept between 400°C and 600°C?   (S4.10)
10 Why is sulphur trioxide not absorbed in pure water? What is used?   (S4.11).

4.15 Questions on the figures and tables

1 Work out the production of sulphuric acid by (a) the UK (b) the USA (c) the USSR if the total world production is 120 million tonnes.   (F4.1)
2 If the UK uses 10 000 tonnes of sulphuric acid each day, how much is used in one year for (a) fertilisers (b) chemicals (c) fibres?   (F4.2)

3 What is the temperature of the gas after the reaction on catalyst bed 2?   (T4.1)
4 What is the final percentage of sulphur dioxide converted to sulphur trioxide?   (T4.1.)
5 Make a list of the products of sulphuric acid and what they are used for.   (T4.2)

4.16 Longer questions

1 Read the optional study in Chapter 3 (Ammonia and nitric acid). Use Le Chatelier's Principle to explain what the best conditions for the Contact Processs should be.
2 Look at Figure 4.3 Explain why sulphur is obtained in this way instead of being mined like coal. Describe how the Frasch Process works.
3 What would happen to industry if the underground deposits of sulphur ran out? What are the alternative sources of supply?
4 Copy Figure 4.4 – Route in a modern plant. Explain in detail what happens at each stage of the process.
5 Explain why an increase in pressure would not have a large effect on the rate of the conversion of sulphur dioxide to sulphur trioxide.
6 The transportation of concentrated sulphuric acid by road tankers is potentially highly dangerous.
   (a) Explain why this is so and suggest the problems that might arise if a serious collision occurred on a main road.
   (b) What other ways of transportation would be safer?

5 Fertilisers: combining plants

5.1 Introduction

Very often an important product is made from two or more different chemicals, which have to be made in separate plants. It is obviously cheaper if these chemicals are made by the same firm, very close to each other. A very good example of this is the I.C.I. Agricultural Division at Billingham, Cleveland (see Figure 5.1).

5.2 Choosing the Billingham site

The chemical industry originally started in the area because it had the following resources:
(1) a source of energy – coal.
(2) raw materials – anhydrite, salt, iron ore, limestone.
(3) a good port.
(4) good roads to other parts of Britain.

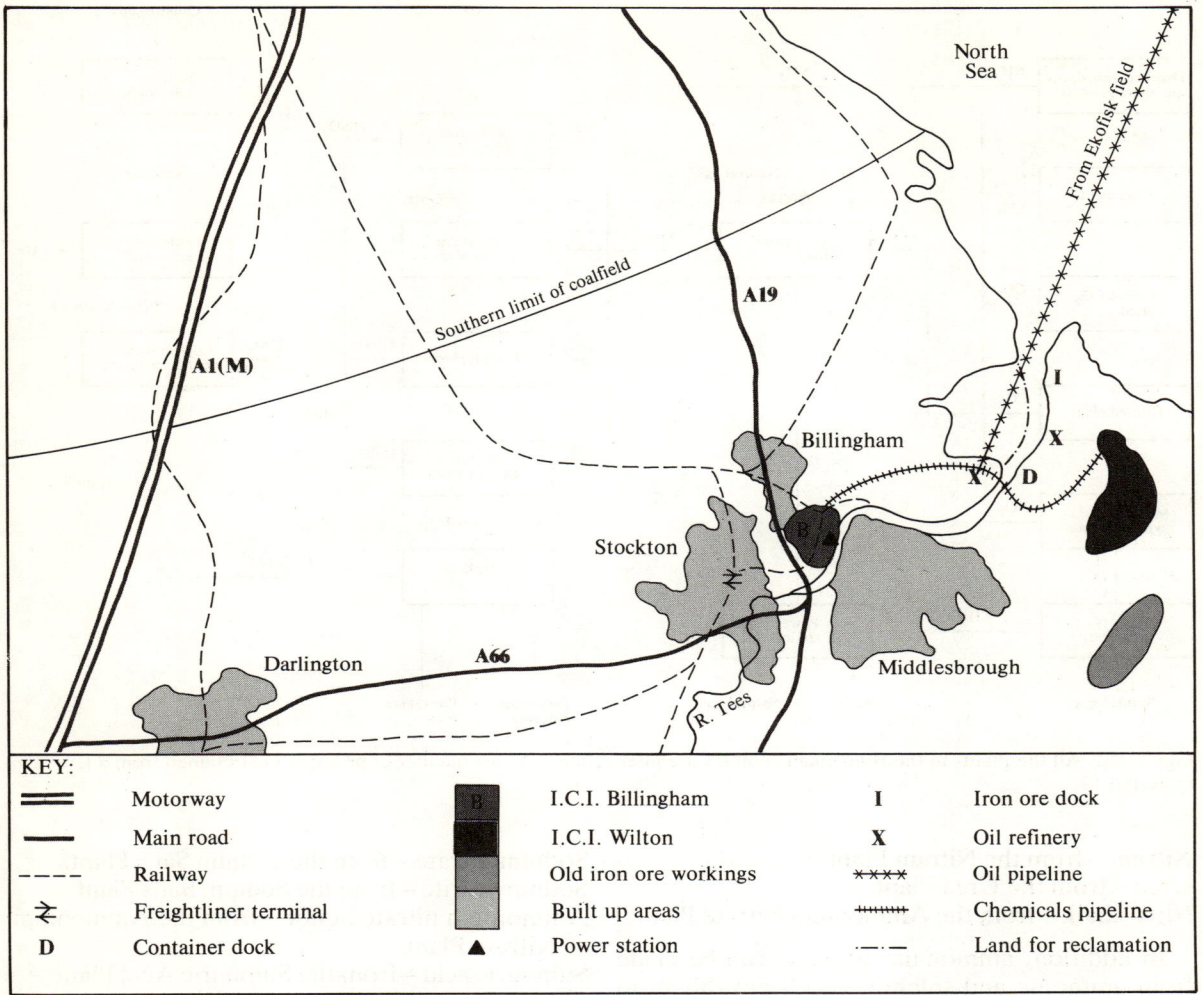

KEY:

═══	Motorway	▨ B	I.C.I. Billingham	**I**	Iron ore dock
────	Main road	▨ W	I.C.I. Wilton	**X**	Oil refinery
- - - -	Railway	▨	Old iron ore workings	××××	Oil pipeline
⇌	Freightliner terminal	▨	Built up areas	┼┼┼┼	Chemicals pipeline
D	Container dock	▲	Power station	-·-·-	Land for reclamation

Figure 5.1. Teesside, showing the important factors relating to the siting of the Billingham fertiliser complex. Other complexes in the same area allow easy transfer of chemicals from one factory to another.

The first railway was built in the area and was soon extended to link up with other ports and industrial areas.

The needs of the area have changed as the raw materials have run out, better sources have been found and new forms of energy and transport are used.

Billingham was developed as a fertiliser plant between the two world wars. It was chosen because the area had built up good access by road, rail and sea, and it had plenty of workers, nearby coal and water and a large new power station. It has developed so that many different plants are inter-linked, using the products of each other and turning a few simple raw materials into many essential chemicals as shown in Figure 5.2.

5.3 Producing fertilisers

The raw materials used:

Natural gas – from the North Sea.
Water – from Teesdale reservoirs.
Air.
Caustic soda – from other I.C.I. factories in Billingham.
Potassium chloride – from the mine at Bowlby, North Yorkshire (partly owned by I.C.I.).
Phosphate rock – from the USA.
Sulphur – from the South of France or the USA.

The fertilisers made:

Compound fertilisers – from the Compound Fertiliser Plant.

27

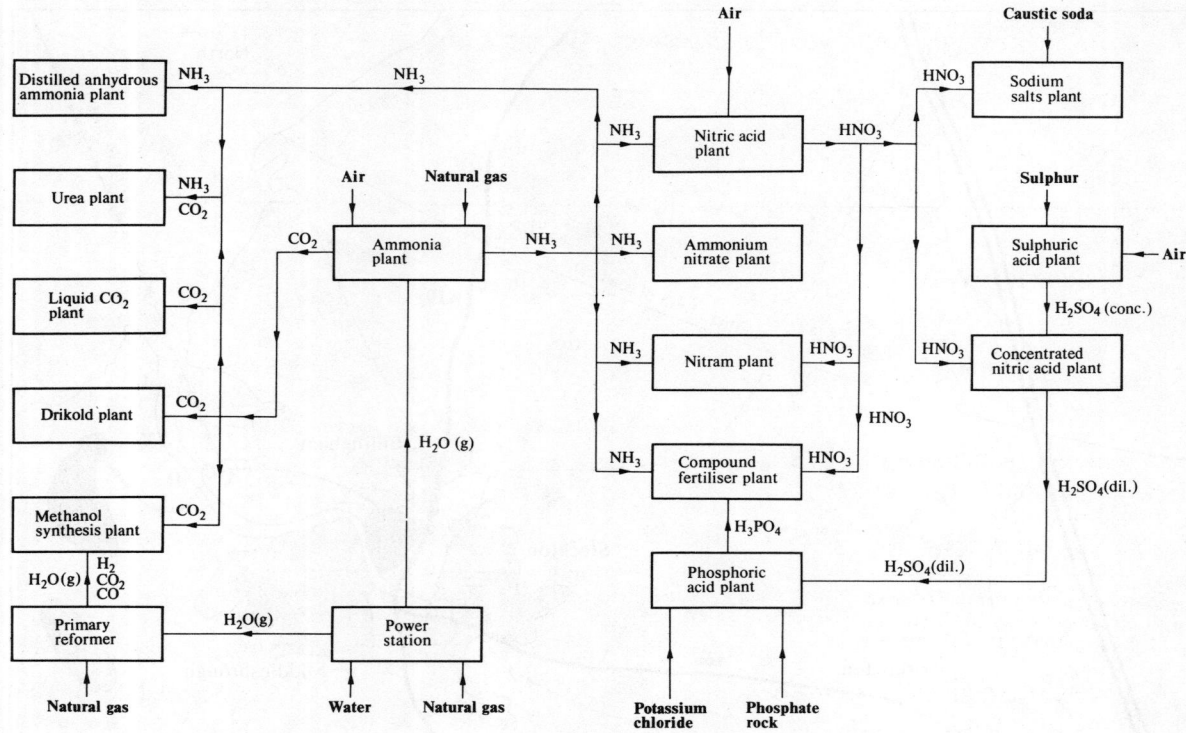

Figure 5.2. All the plants in the Billingham complex are inter-related. A vast number of products are obtained from a few raw materials.

Nitram – from the Nitram Plant.
Urea – from the Urea Plant.
Nitro chalk – from the Ammonium Nitrate Plant.

In addition, ammonium sulphate can be made from ammonia and sulphuric acid, but this is no longer produced at Billingham. Sodium nitrate and ammonia can also be used as fertilisers.

5.4 Other products

Products used for purposes other than fertilisers:

Electricity — from the Power Station.
Steam – from the Power Station.
Methanol – from the Methanol Synthesis Plant.
Ammonia solution – from the Ammonia Plant.
Hydrogen – from the Ammonia Plant.
Urea – from the Urea Plant.
Drikold (solid CO_2) – from the Drikold Plant.
Liquid CO_2 – from the Liquid CO_2 Plant.
Liquid ammonia – from the Distilled Anhydrous Ammonia Plant.
Dilute nitric acid – from the Nitric Acid Plant.
Concentrated nitric acid – from the Concentrated Nitric Acid Plant.

Sodium nitrate – from the Sodium Salts Plant.
Sodium nitrite – from the Sodium Salts Plant.
Ammonium nitrate liquor – from the Ammonium Nitrate Plant.
Sulphuric acid – from the Sulphuric Acid Plant.

Many of these are used in other divisions of I.C.I. at Billingham, Wilton and other areas of the UK.

5.5 N, P and K

The importance of fertilisers has already been mentioned in the chapter on Ammonia. It has been found that the most important elements needed for fertilisers are nitrogen (N), phosphorus (P) and potassium (K). The amounts of these elements present in fertilisers are shown in Table 5.1.

5.6 Questions on the text sections

1 Why did the chemical industries start in Teesside? (S5.2)
2 What raw materials are used in the Agri-

Table 5.1 Percentages of nitrogen (N), phosphorus (P) and potassium (K) in fertilisers.

Fertiliser	Formula	%N	%P	%K
Compound fertilisers*	NH_4NO_3	15	7	18
	+			
	$NH_4H_2PO_4$	23	5	19
	+			
	KCl	9	11	21
Nitro-chalk	$NH_4NO_3 + CaCO_3$	25		
Nitram	NH_4NO_3	35		
Urea	$(NH_2)_2CO$	47		
Ammonium sulphate	$(NH_4)_2SO_4$	21		
Ammonia solution	NH_4OH	40		
Sodium nitrate	$NaNO_3$	16		
Calcium superphosphate	$Ca(H_2PO_4)_2$		20	

* There are three different compound fertilisers. They all contain the same compounds, but in different proportions. They have the three compositions shown.

cultural Division and where do they come from? (S5.3)

3 Which fertilisers are made at Billingham? (S5.3)

4 What other products are obtained as well as fertilisers? (S5.4)

5 Which are the three most important elements found in fertilisers? (S5.5)

5.7 Questions on the figures and tables

1 What connects the I.C.I. complexes at Billingham and Wilton? (F5.1)

2 Which methods of transport are available near Billingham? (F5.1)

3 Make a list of all the plants in the Agricultural Division (F5.2)

4 Which plants use steam ($H_2O(g)$)? (F5.2)

5 Which plants use (a) carbon dioxide (b) ammonia, from the Ammonia Plant? (F5.2)

6 Which plants use nitric acid? (F5.2)

7 Which chemicals does the Methanol Synthesis Plant use? (F5.2)

8 Which fertilisers contain N, P and K? (T.5.1)

5.8 Longer questions

1 Explain why dilute sulphuric acid is obtained from the Concentrated Nitric Acid Plant.

2 The Teesside factories need a good supply of water, which can only be obtained by building a new reservoir in Upper Teesdale. Unfortunately this will mean drowning some rare alpine flowers.

Either Prepare a summary paper for the Department of the Environment explaining the importance to the British economy of building the reservoir.

Or Form a discussion group debating the issue from the points of view of an Environmental Protection Group and an Industrial Management Group.

3 You have been asked by a large chemical firm to look for a site for a new fertiliser complex. Explain, with reasons, what you would look for when selecting a suitable site.

4 A small explosion has occurred at the Nitric Acid Plant in a fertiliser complex. Explain how this could affect (a) the other plants in the complex (b) the people living in the area.

5 A compound fertiliser is made which contains ammonium nitrate, ammonium dihydrogen phosphate and potassium chloride in the ratio:

1 mole	NH_4NO_3	(1 mole = molecular
2 moles	$NH_4H_2PO_4$	mass in grammes)
2 moles	KCl	

Calculate the percentage of nitrogen, phosphorus and potassium in the fertiliser. (Relative atomic masses: H = 1, N = 14, O = 16, P = 31, Cl = 35.5, K = 39.)

6 Iron and steel

6.1 Introduction

Try to imagine a world without iron and steel and you will quickly realise that we would have made little technological progress over the last five hundred years. Iron and steel products are everywhere around us: in transport, building construction and in agriculture. They are used in the chemical and engineering industries, in the home and in our leisure activities. Machines made of steel process most of the other materials we use and the products we manufacture. Indeed, iron and steel must rank as two of the most important materials that man produces.

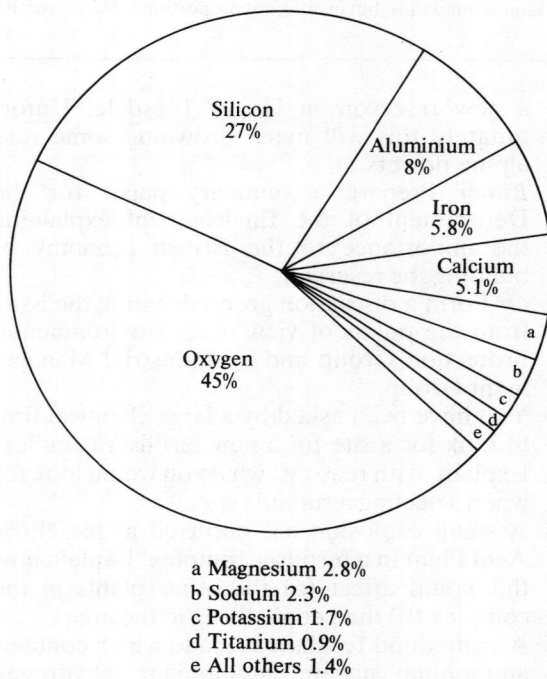

a Magnesium 2.8%
b Sodium 2.3%
c Potassium 1.7%
d Titanium 0.9%
e All others 1.4%

Figure 6.1. The occurrence of elements in the earth's crust.

6.2 Occurrence of iron

Iron is the second most abundant metal and the fourth most abundant element in the earth's crust (see Figure 6.1). It is found in most parts of the world, usually in the form of oxides, carbonates, sulphates and silicates. However, most of the iron produced comes from the oxide ores, **haematite,** Fe_2O_3 and **magnetite,** Fe_3O_4. Most of the ores used today contain between 20% and 70% iron, the remaining percentage being mainly sand and clay. UK ores are of poorer quality containing about 28% iron and deposits are found in Cumberland, Oxfordshire, Leicestershire, Lincolnshire and Northamptonshire. Higher quality ores are preferred these days and are found mainly in Scandinavia, Australia, the Americas, North Africa and Russia. The ore is obtained by quarrying or open cast mining.

6.3 Historical development

3000 years ago it was known that when iron oxide was heated in a charcoal fire, a spongy solid mass of iron was obtained that could be beaten into shape and used for tools and weapons. Eventually primitive furnaces were built into the hillsides, facing the prevailing wind. The furnace with its charge of charcoal and iron ore was left to burn for 3 or 4 days. After the fire had gone out, solid lumps of iron were removed and the process was started up again. These were known as 'batch furnaces'.

The Staffordshire hills around Cannock and the Weald of Kent were important sites of iron foundries in the Middle Ages. It was during the 1300s and 1400s that the crude furnaces were converted to take a constant blast of air by using bellows often operated by a water wheel. This forced air blast allowed higher temperatures in the furnaces causing the iron to melt and by this time the iron founders had learned how to handle and cast the molten iron into moulds.

6.4 The industrial revolution

Up until the 1600s the fuel used in the furnaces was charcoal. Dud Dudley pioneered the use of coal and coke as a fuel but it was not until the 1700s that coke alone was used. Abraham Darby at Coalbrookdale in Shropshire began using coke in the smelting of iron in the blast furnace.

Through the 1800s the blast furnaces grew in size and complexity. A Shropshire furnace in the eighteenth century made about 20 tonnes of iron each week. A modern blast furnace can produce up to 10 000 tonnes in a day!

6.5 Steel making

Steel making, by comparison with iron making, is a very young industry. During the last 100 years the industry has undergone many new developments from the early processes of Siemens-Martin open hearth furnaces, 1851 and the Henry Bessemer Process (which is the basis of the Basic Oxygen Process) 1856, improved by Gilchrist and Thomas in 1879. The world yearly production of steel in the 1860s was about 300 000 tonnes. Compare that with the 746 million tonnes produced in 1979.

The Iron and Steel Industry can today be divided into:

Iron making: the production of iron from iron ore.

Steel making: refining of the 'impure' iron to produce various grades of steel.

6.6 The chemistry of the process

The smelting of iron ore is essentially the removal of oxygen from iron oxide. This process is known as **reduction** and requires a **reducing agent** and heat to carry it out, e.g.

$$Fe_2O_3(s) + 3CO(g) \longrightarrow 2Fe(l) + 3CO_2(g)$$
iron ore   carbon   iron   carbon
          monoxide   dioxide

The reducing agent, carbon monoxide, can be made, in two stages:

(1) burning coke (carbon) in air to give carbon dioxide:

$$C(s) + O_2(g) \longrightarrow CO_2(g) + heat$$
carbon   oxygen   carbon
                dioxide

(2) reacting the carbon dioxide produced with more coke to give carbon monoxide:

$$CO_2(g) + C(s) \longrightarrow 2CO(g)$$
carbon   carbon   carbon
dioxide   monoxide

The chemistry of steel making is basically the removal of unwanted elements in the iron metal by making them combine with oxygen. This is known as **oxidation** and requires an **oxidising agent** and heat to carry it out, e.g.

$$C(s) + O_2(g) \longrightarrow CO_2(g)$$
+ other   oxidising   + other
impurities   agent   oxides

6.7 Location of the industry

The raw materials for iron making are heavy and bulky and transport has always been one of the problems of the industry. This led during the eighteenth and nineteenth centuries to the industry developing in areas where rich iron ore, charcoal (forests) and limestone were available locally. As transport facilities improved (e.g. railways, canals) the demand for steel grew and coke began to replace charcoal as a fuel, the industry moved into areas where good quality smelting coal and larger deposits of iron ore were available. Although UK ore was generally lower grade, there was plenty of it, so large plants were built on the orefields, e.g. in Northamptonshire and Lincolnshire.

Table 6.1 The quantities of home and imported iron ore used between 1950 and 1977.

Year	Home ores (tonnes)	Imported ores (tonnes)
1950	12 963 000	8 350 900
1955	16 175 100	12 812 000
1960	17 087 400	17 335 000
1965	15 414 700	18 211 000
1970	12 103 500	19 262 350
1975	4 709 000	16 005 000
1977	3 973 000	16 379 000

Modern developments in the industry required large quantities of high grade ore (60% Fe) so nowadays at least half the iron ore requirements of the UK come from abroad, e.g. Sweden, Canada, Mauritania and Venezuela (see Table 6.1). Massive ship freighters capable of carrying over 300 000 tonnes of ore have made it economical to import ore. These now require new ore terminals large enough to cope with their huge size. This has led the industry to develop its large existing plants near the coastal ore terminals in Scotland, South Wales, Humberside and Teesside (see Figure 6.2). Inland transport has been kept to a minimum, and many of the old works which were based on home

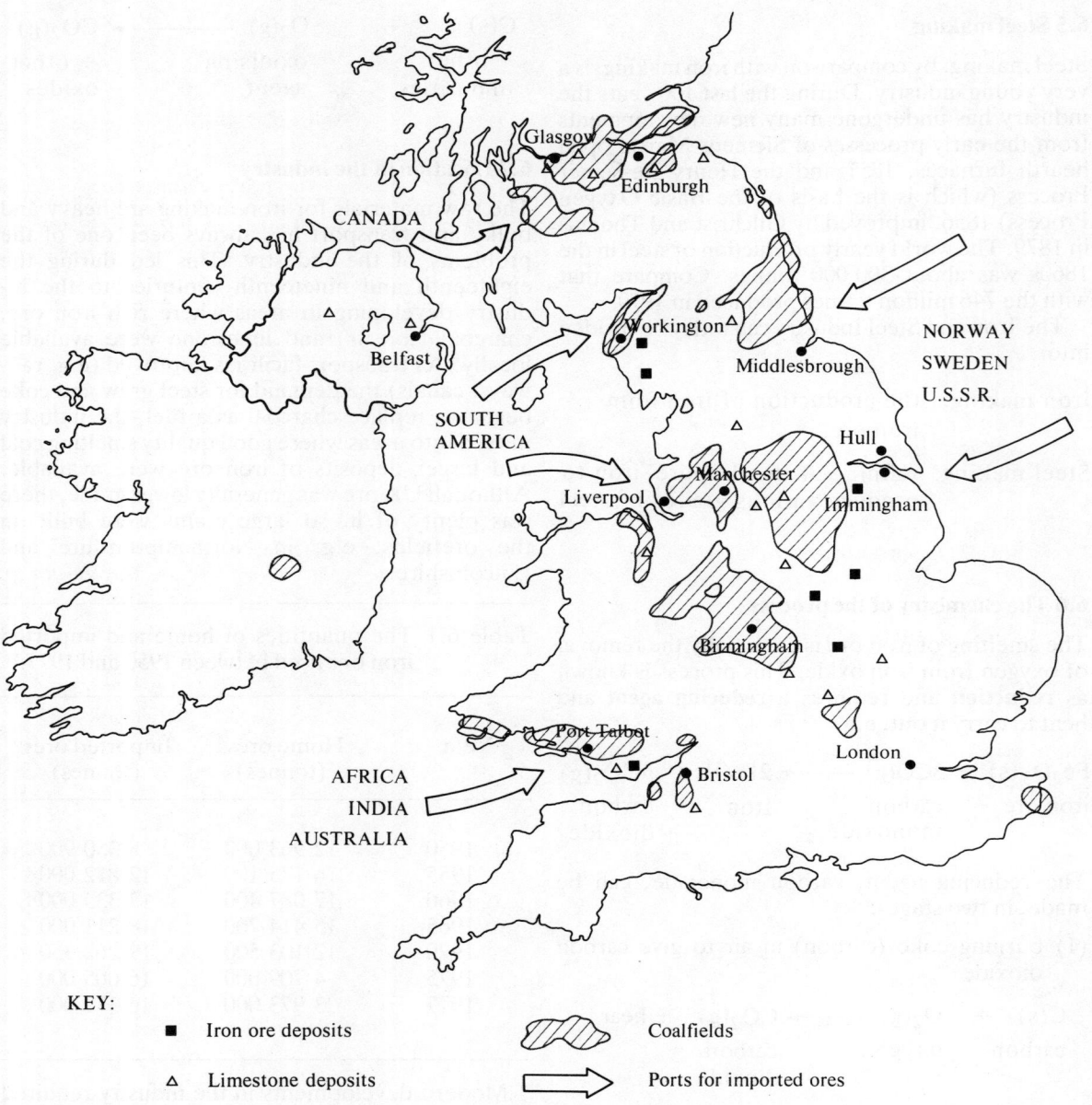

Iron ore deposits   ▨   Coalfields

△   Limestone deposits   ⇨   Ports for imported ores

Figure 6.2. The map shows the location of the raw materials required for the making of iron and steel. It also shows the main ports to which imported ores and smelting coal are brought.

orefields have closed. Furthermore a good deal of high quality smelting coal is now also imported.

6.8 The modern manufacturing process

The processes of iron and steel making are nowadays usually carried on in large integrated iron and steel plants (see Figure 6.3).

6.9 Iron making

The smelting of iron ore is carried out in a **blast furnace** (see Figure 6.4). The modern blast furnace is made of steel and can be up to 70 m in height. It consists of a steel cylinder about 30 m high, with a base or hearth about 10 m in diameter. The cylinder is lined with fire-proof bricks on the inside and on the top it is fitted with a system for loading

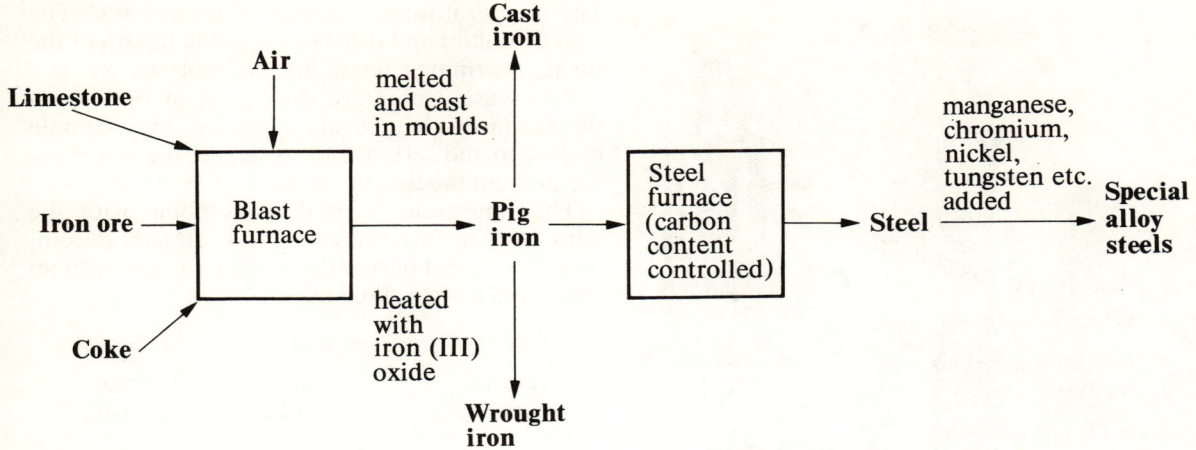

Figure 6.3. The route in a modern integrated iron and steel plant.

the raw materials, and the outlet pipes for the waste gases. At the bottom of the cylinder are a series of nozzles called tuyères through which the hot air is blown in from the hot blast stoves.

The raw materials, known as the charge, consist of a mixture of **iron ore** (in the form of pellets or 'sinter' – a mixture of ore, limestone and coke roasted together in a sinter plant), **coke** (specially prepared from high quality coal) and **limestone** are charged through the top of the furnace. Hot air is blown in through the tuyères at the bottom and provides the heat needed to melt the charge and carry out the various chemical reactions. The process can be considered in three stages:

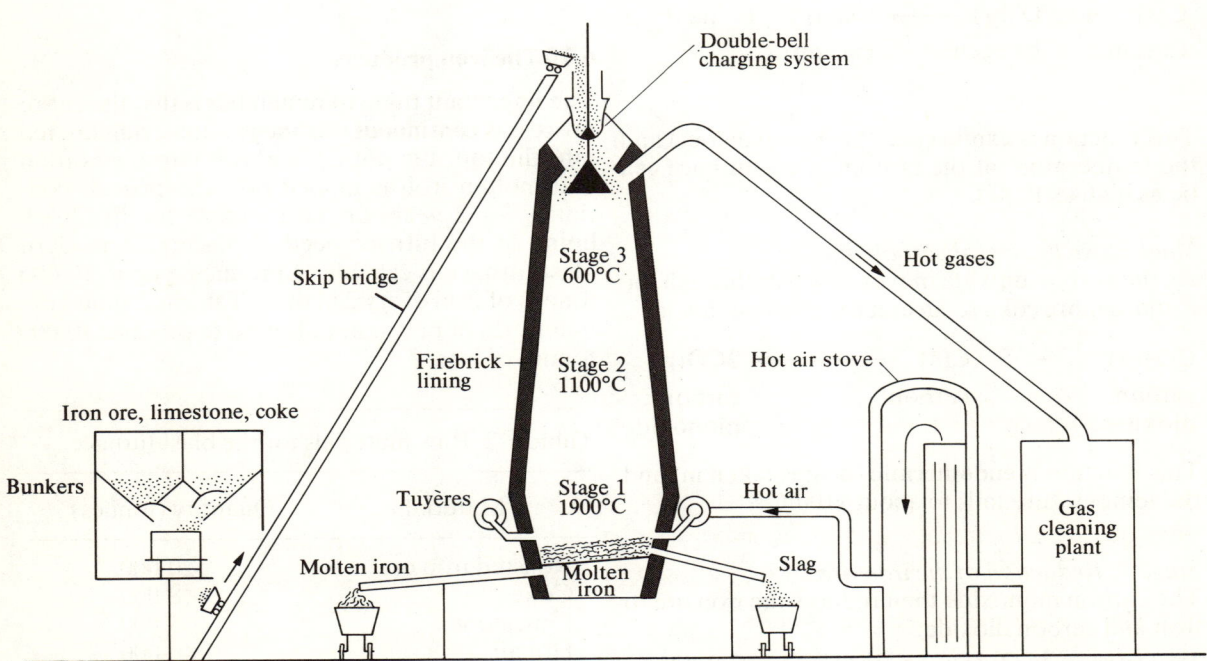

Figure 6.4. Smelting iron ore in the blast furnace is a continuous process carrying on for may be two years and stopping only when the fire brick lining of the furnace is renewed.

33

Blast furnace at Llanwern works, Newport.

The temperature at this stage is about **600°C**. The iron is molten and drips down to the hearth of the furnace forming a heavy layer of molten iron.

The waste gases are drawn off at the top of the furnace, cleaned and, because they contain hydrogen and carbon monoxide, are then used as a fuel to heat the hot air stoves.

The limestone is used to combine with the impurities in the iron ore. It has already decomposed in the hot part of the furnace to form calcium oxide and carbon dioxide.

$$CaCO_3(s) \longrightarrow CaO(s) + CO_2(g)$$

calcium	calcium	carbon
carbonate	oxide	dioxide

The calcium oxide then combines with the silica (sand) impurities in the ore to form calcium silicate (slag).

$$CaO(s) + SiO_2(s) \longrightarrow CaSiO_3(l)$$

calcium	silicon	calcium
oxide	dioxide	silicate

The calcium silicate is in the form of a liquid slag and falls to the hearth where it floats on top of the molten iron.

6.10 The iron products

The important thing to remember is that the entire process is **continuous**. As the raw materials are fed into the top, the hot air is blown into the bottom and molten iron is tapped off. The process continues for 2 years or more before the fire brick lining of the furnace needs replacing. A modern blast furnace is capable of producing up to 10 000 tonnes of iron every 24 hours. Table 6.2 shows the quantities of raw materials used to produce 10 000 tonnes of iron.

Table 6.2 Raw materials for the blast furnace.

Material	Quantity (tonnes)
Roasted iron ore	20 000
Coke	8 000
Limestone	5 000
Hot air	40 000
Iron produced	10 000

Stage 1: Oxidation or combustion of carbon
In the bottom part of the furnace the coke burns in the hot air to form carbon dioxide.

$$C(s) + O_2(g) \longrightarrow CO_2(g) + heat$$

carbon	oxygen	carbon
		dioxide

This reaction is **exothermic** (heat is given out) and the temperature at the bottom of the furnace can be as high as 1900°C.

Stage 2: Reduction of carbon dioxide
As the carbon dioxide moves upwards through the furnace more coke reduces it to carbon monoxide.

$$CO_2(g) + C(s) \longrightarrow 2CO(g)$$

carbon	carbon	carbon
dioxide		monoxide

This reaction is **endothermic** (heat is taken in) and the temperature falls to about 1100°C.

Stage 3: Reduction of the iron ore
The carbon monoxide then reduces the iron ore to iron and carbon dioxide.

$$Fe_2O_3(s) + 3CO(g) \longrightarrow 2Fe(l) + 3CO_2(g)$$

iron(III)	carbon	iron	carbon
oxide	monoxide		dioxide

Table 6.3 The properties and uses of cast and wrought iron.

Type	Properties	Uses
Cast iron	Contains up to 4% carbon. Brittle. High compression strength.	Car engine blocks, gas stoves, bunsen burner bases, pipes, machinery, man hole covers.
Wrought iron	Purest form of iron (99% iron). Soft, bends easily. Can be worked without breaking	Nails, bolts, chains, garden gates. (The Eiffel Tower was made from 7300 tonnes of wrought iron girders.)

When sufficient iron has been produced in the furnace, both the slag and the iron are tapped off through separate holes. The slag is used for road making, cement and building blocks.

Some of the iron is run off into moulds to form ingots of **cast iron** (known also as **'pig iron'**). Some of the cast iron is then converted to **wrought** iron by a process called **'puddling',** in which it is first partly melted, and then stirred so that some of the carbon is removed by oxidation. The hot iron is then squeezed and hammered between giant rollers and hammers to 'squeeze' out any further carbon to the surface, when it is ejected as a scale. Table 6.3 lists the properties and some uses of cast and wrought iron.

Most of the iron is nowadays taken straight to the steel furnaces in giant ladles whilst still molten, and mixed with scrap iron for conversion to various grades of steel.

6.11 Steel making

The molten iron tapped from the furnace contains about 7% impurities, the main ones being carbon, silicon, manganese, phosphorus and sulphur. The process of steel making involves reducing or even eliminating these, controlling the percentages of carbon, and then adding carefully controlled amounts of other elements to produce the desired types of steel. In general if iron contains less than about 1.7% of carbon it is classed as a steel. There are soft, hard, springy, special electric steels and many alloys where iron is mixed with other metals to form special and stainless steels.

The next three sections deal with the three main processes by which steel is made.

6.12 Open-hearth Process

The raw materials, **scrap**, **molten iron**, and **limestone**, are charged into a shallow open furnace (Figure 6.5) and exposed to the flames of burning oil or gas. Oxygen is injected into the furnace to assist the oxidation of the unwanted elements

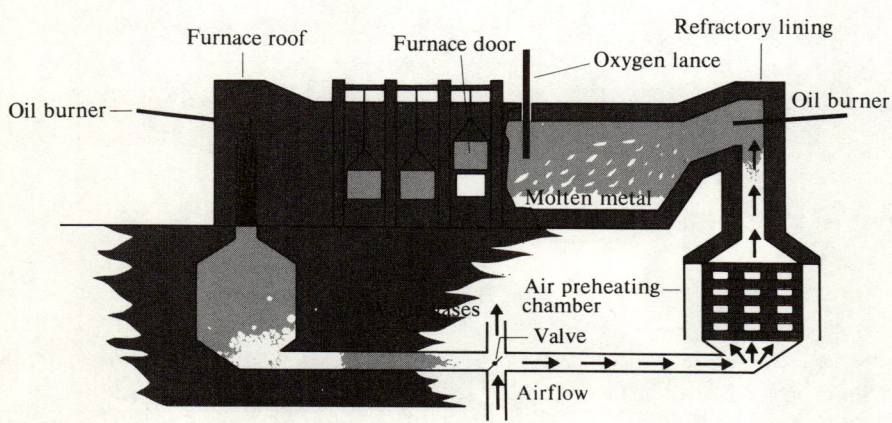

Figure 6.5. The open-hearth furnace.

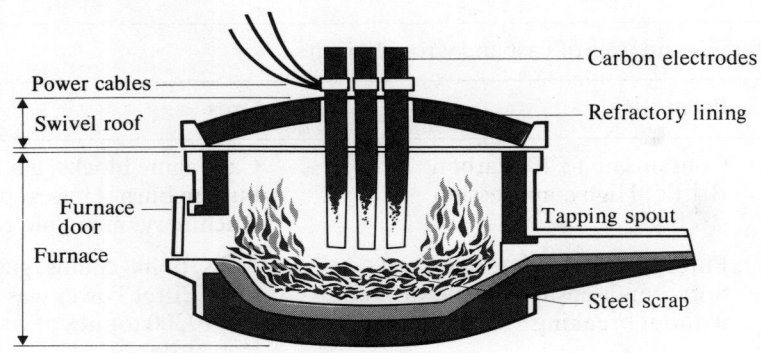

Figure 6.6. The electric arc furnace.

which combine with the limestone to form a slag. The slag floats on the molten steel and can easily be separated.

The carbon content is either reduced to the required percentage or is removed almost completely, and the calculated amount is added.

This process produces about 350 tonnes of steel in about 10 hours and is regarded as being slow. It is gradually being replaced by the following two processes.

6.13 The Electric Arc Process
This process uses scrap metal and, unlike other steel making processes, hot metal from the blast furnace is not used (see Figure 6.6).

The furnace consists of a circular bath and a movable roof through which huge graphite electrodes pass and can be raised or lowered. **Steel scrap** is charged into the furnace and a large electric current of up to 40 000 amps is passed through the electrodes, causing arcing and raising

Tapping an electric arc furnace.

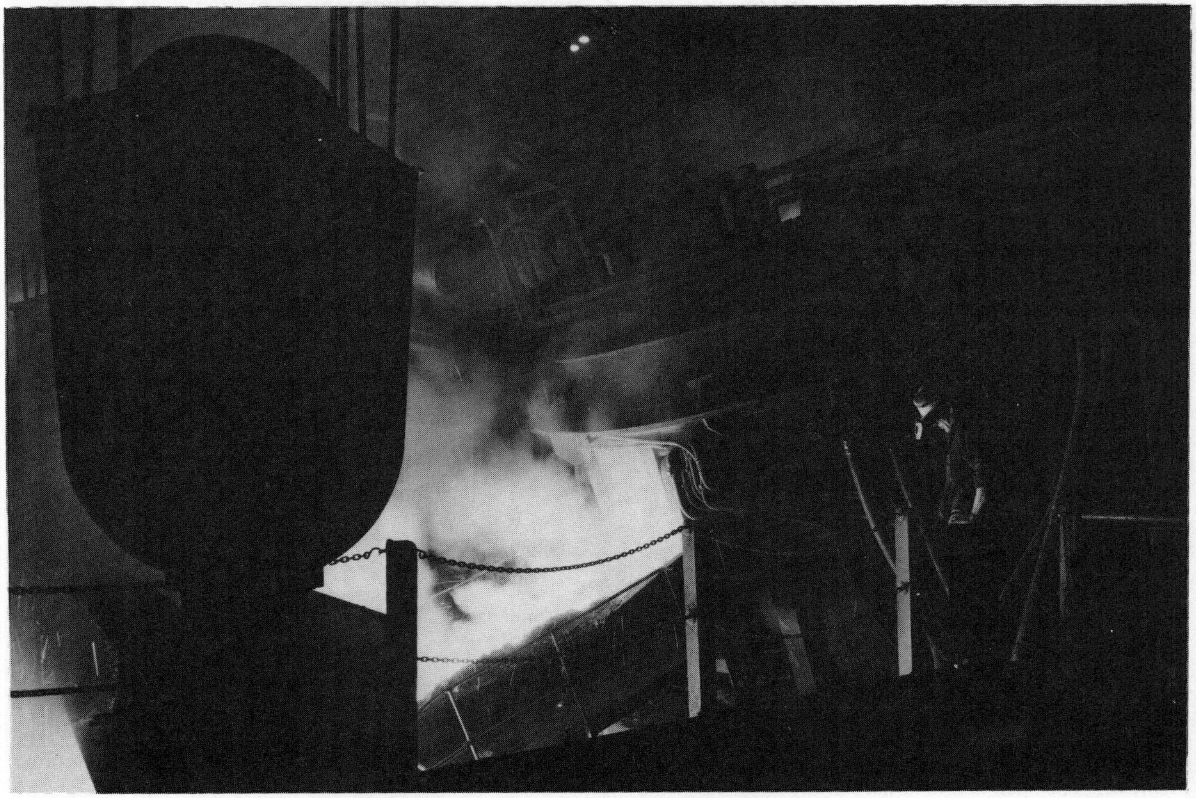

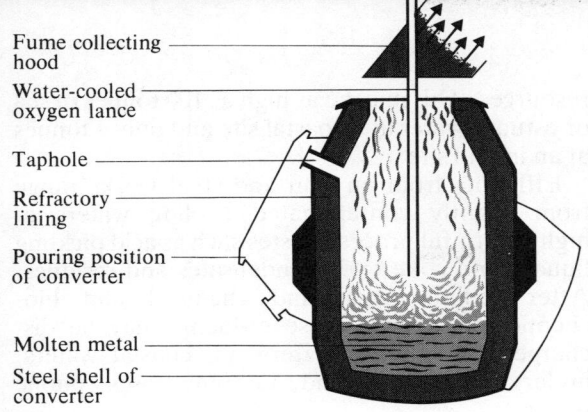

Fume collecting hood
Water-cooled oxygen lance
Taphole
Refractory lining
Pouring position of converter
Molten metal
Steel shell of converter

Figure 6.7. The basic oxygen furnace.

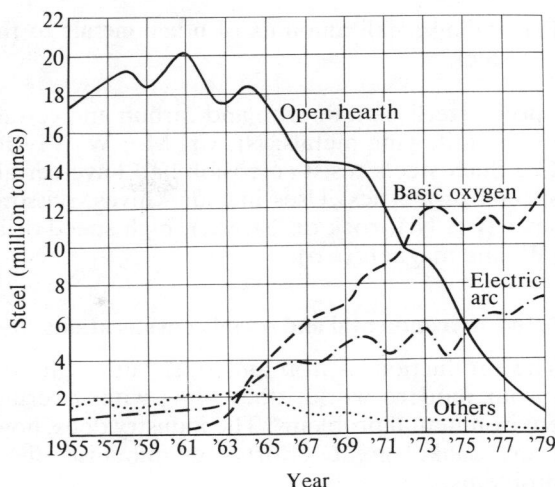

Figure 6.8. The annual production of steel in the UK by the various processes between 1955 and 1979.

the temperature up to 3400°C, which melts the scrap. **Limestone** and **iron oxide** are added and combine with the impurities to form a liquid slag. When the correct composition of the steel has been achieved the furnace is tilted and first the slag is poured off, followed by the steel. The large electric furnaces can produce about 150 tonnes of steel in about 4 hours.

6.14 The Basic Oxygen Process

This process is fast becoming the major method of producing steel (see Figure 6.7). The furnace or converter is tilted so that it can first be charged with **scrap iron** (30%) and then **molten iron** (70%). A water-cooled copper lance is lowered into the furnace and pure oxygen is blown through it onto the metal surface at high speed and high pressure. The converter rotates during the oxygen blow and the impurities of carbon and phosphorus are blown out of the molten iron as gaseous oxides. During the blow limestone is added which combines with the other impurities forming a slag. The converter is first tilted and the molten steel tapped off, then tilted the other way and the slag removed. Modern furnaces can produce about 350 tonnes of steel in 40 minutes.

Figure 6.8 shows the annual production of steel in the UK by the various processes between 1955 and 1979.

6.15 Properties and uses of steels

The properties and uses of steel can be varied by:

(1) controlling the amount of **carbon** in the steel (see Table 6.4).
(2) subjecting the steel to the heat treatment processes of **hardening** and **tempering**. It is made hard but brittle by making it red hot and then cooling it quickly in oil or water. By careful heating at about 300°C (**tempering**) steel loses its brittleness but retains its hardness.

Table 6.4 The properties and uses of different types of steel.

Type	% Carbon	Uses	Properties	
Soft steels	up to 0.15	Wire, rivets, cans, car bodies	Soft, strong, ductile	
Mild steels	0.15 – 0.25	Machines, girders, ship's plate	Strong, ductile	Hardness increases Strength decreases Ductility decreases
Medium steels	0.25 – 0.50	Springs, rails	Hard, not so strong, low ductility	
Hard steels	0.50 – 1.70	Tools, scissors, knives, razors	Hard, brittle, low ductility	

(3) adding small amounts of **other metals** to the steel.

Special steels contain iron and carbon and certain of the following metals: Ni, Cr, Mo, W, V, Mn, Co. Such steels resist corrosion and have certain special properties. Uses include knives, scissors, saws (Cr, Ni), rock drills (Mn), high speed drills (W) and magnets (Co).

6.16 Environmental and social considerations

The production of iron and steel, although vital to our modern world, does bring with it certain environmental problems. The industry does, however, make great efforts to minimise these problems.

Iron ore which is on or near the surface can be removed by quarrying, e.g. this was the case at the Frodingham iron ore deposits in Lincolnshire. Quarrying causes severe devastation of the landscape and returning it to an acceptable state is a slow process.

Air pollution

The main sources of air pollution arise from:

(1) sulphur gases, mostly **sulphur dioxide** produced during the combustion of fuels (oil and coal), sintering of the ore and the use of limestone, and as an impurity during the steel making process.
(2) black smoke from the coal burning plant.
(3) grit and dust from ash and scale, slag, sinter and iron ore.
(4) very fine particles of iron oxide suspended in the waste gases during oxygen steel making, giving rise to reddish brown fumes.

Smoke and gases are usually dispersed through tall chimneys. Dust and grit is filtered out. Attention is now being given to removing sulphur from fuels and waste gases before it is emitted (usually as sulphur dioxide) to the atmosphere.

In newer plants air pollution is carefully controlled; however, some older plants are in need of improvement.

Water pollution

The industry uses vast amounts of water (about 200 tonnes of water are needed to produce 1 tonne of steel) much of which is recirculated and reused. In fact, on average about 37 tonnes of water per tonne of steel are actually consumed from water

resources. This may be as high as 100 tonnes of sea or estuary water at a coastal site and only 4 tonnes at an inland site.

Effluents from an iron and steel works range from slightly contaminated cooling waters to highly harmful process wastes such as acid pickling liquors and coke oven condensates and residues. After sedimentation and chemical and bio-chemical treatment these effluents may be discharged into inland waterways, coastal waters, underground strata and, in some cases, public sewers.

The industry has made efforts to reduce water pollution, but problems still remain, e.g. the Ebbw Vale works heavily polluted the River Ebbw. Fortunately installations now fitted there helped to resolve that problem.

As the industry becomes concentrated in larger manufacturing units the pollution problems are likely to become more acute and the need for even greater control will be important.

Corrosion

Unfortunately iron and steel rust away when exposed to air and water. In most cases this is unsightly, and in some cases dangerous. Several hundred thousand tonnes of iron and steel turn to rust every year in the world. It has been estimated that of the 1800 million tonnes of steel produced between 1890 and 1923 almost half of it has been lost through rusting. Corrosion and corrosion protection cost industry several £1000 million per year.

6.17 The future of steel making in Britain

Only a few private firms make steel. The vast majority is produced by the British Steel Corporation, a nationalised industry. The B.S.C. returned losses of £443 million and £309 million in 1978 and 1979. This led to the company streamlining its steel making programme. The industry has therefore recently been concentrated into a small number of very large integrated plants, chiefly in South Wales, Teesside, Humberside, and Clydeside. All of these plants are close to port facilities, so high grade imported ores and coking coal can be used. These developments have resulted in the closure of many older plants and the reduction of the workforce in the whole industry by about 50 000 (30%).

Certain areas of the country have been badly hit, particularly South Wales, the Midlands and the North-east of England. Corby was a village of 1500 people until it became a steel town. Local deposits of iron ore and coal led the B.S.C. to set up a large iron and steel works helped by a £7 million investment grant and other inducements by the Government. Eventually Corby new town developed with a population of 55 000, half of whom were employed by, or depended upon, the steel works for their livelihood.

In 1979 it was decided that it was no longer economical to make steel at Corby. Local iron ore is low grade and it is costly on fuel to extract the iron. The plant closed down in 1980 resulting in a very large number of redundancies and high unemployment in Corby. Other companies which depend on the iron and steel industry have also been seriously affected.

On the technical side research is being carried on to establish an improved steel making process route using low grade iron ore from Saudi Arabia and also to develop energy conservation schemes for key iron and steel making operations. Further research continues on the improvement of the continuous iron and steel making process in which the metal is at all stages kept in the molten state with continuous transfer from blast furnace to steel furnace, continuous removal of slag and sulphur and continuous and automatic control of metal analysis and temperatures.

6.18 Questions on the text sections

1 In which countries are high quality iron ores found?   (S6.2)
2 Give the names and formulae of the two most common oxide ores of iron.   (S6.2)
3 Can you think of any reasons why the Cannock hills in Staffordshire and the Weald of Kent were important sites for early iron founders?   (S6.3)
4 Name the compound formed when hot air is passed through coke.   (S6.6)
5 Why is the reaction in question 4 important in reducing fuel costs for the process?   (S6.9)
6 (a) Which compound reduces iron ore in the blast furnace?
   (b) How is this compound formed?   (S6.6)
7 Write word and symbol equations for the reduction of iron(III) oxide.   (S6.9)
8 Calcium carbonate (limestone) is decomposed in the blast furnace. Write word and symbol equations for the decomposition.   (S6.9)
9 Why is limestone added to the furnace?   (S6.9)

10 What are the waste gases drawn off from the top of the furnace used for?   (S6.9)
11 Give two uses for pig or cast iron other than for steelmaking.   (S6.10)
12 Give one use for the slag which is produced.   (S6.10)
13 Calculate the mass of iron which should be obtained if 80 tonnes of iron(III) oxide (Fe_2O_3) were completely reduced. (Relative atomic masses O = 16, Fe = 56.) (S6.6).
14 Name two impurities present in iron obtained from the blast furnace.   (S6.11)
15 Name the raw materials used in the Basic Oxygen Process. What is the purpose of the copper lance? Which two impurities are removed as gaseous oxides? What is the purpose of adding limestone?   (S6.14).

6.19 Questions on the figures and tables

1 Refer to the map and indicate suitable sites for iron and steel works in the nineteenth century. Bearing in mind that nowadays most of our iron ore is imported, suggest suitable locations for modern iron and steel works.   (F6.2)
2 What do the figures in the table tell you about the trends in the use of home iron ores compared with imported ores? Give reasons to explain why you think this trend is as it is. (T6.1.)
3 What is the difference between cast iron and wrought iron? Make a list of the uses of each. (T6.3)
4 (a) Work out the total tonnage of steel produced in 1970 and 1979.
   (b) Can you think of any reason why UK steel production in 1970 was greater than it was in 1979?
   (c) Comment on the trends in the production of steel by the three main processes between 1955 and 1979.   (F6.8)

6.20 Longer questions

1 Read carefully the section on the properties and uses of steel and make a summary.
2 Write a short account highlighting the pollution problems associated with the iron and steel industry.
3 Discuss the advantages and disadvantages of the British Steel Corporation's plans to develop large integrated plants at coastal sites.
4 Write a short account of the problems associated with the corrosion of steel. Try to find some methods by which it can be prevented.

5 A suggestion for debate:

Form into groups of three or four. Some groups should represent members of the Newtown Joint Action Committee (J.A.C.) made up from Trade Unionists, Local Council members, local businessmen and the other groups can represent members of the B.S.C. Board. The groups representing Newtown J.A.C. should put forward the case, giving reasoned arguments, for keeping the iron and steel plant in operation for the good of the community even though it may be unprofitable. They might also like to suggest ways in which they think the profitability of the plant could be improved. The groups representing the B.S.C. should give reasons why the plant must be closed, outlining alternative forms of employment for the work force and support for the community of Newtown.

7 Aluminium

7.1 Introduction

Discovered little more than 150 years ago and produced commercially for about half that time, aluminium today ranks second only to iron and steel in its service to mankind.

It is light, strong, durable and a good conductor of heat and electricity. The metal can be easily rolled, forged, sawn and slit. It can also be shaped by extrusion, i.e. forced through dies of different shapes. Aluminium can be drawn into wires so thin that one stretched around the world would only weigh a hundred or so kilogrammes. In addition it has the advantage of being able to resist corrosion by forming a protective film of aluminium oxide on its surface. Aluminium is indeed a versatile metal, its uses ranging from kitchen foil to Concorde.

7.2 Occurrence of aluminium

Aluminium is the most common metal on earth. A handful of soils from your garden will probably contain a good quantity of the metal since it forms about $\frac{1}{12}$ of the earth's crust (see Figure 6.1). Aluminium only occurs combined with other elements, mainly oxygen. It is found naturally as **bauxite**, $Al_2O_3.3H_2O$, which is purified to form **alumina**, Al_2O_3. Commercial grade bauxite contains at least:

40% alumina (chemically combined with 15–32% water to form a hydrated oxide)
7% sand ⎫
20% iron oxide ⎬ impurities
4% titanium oxide ⎭

Bauxite deposits are usually found near the surface in layers or pockets about 1–12 m or more thick under a cover of $\frac{1}{2}$ m up to 60 m. This allows the ore to be obtained by open pit mining.

The world's major deposits of bauxite are found in tropical or sub-tropical regions, the main suppliers being Australia and Guinea (accounting for almost half the world deposits). Jamaica, Surinam and Guyana, Indonesia, India and the Cameroons also possess large deposits. Recent discoveries in Brazil may rival the massive deposits in Australia and Guinea. Table 7.1 shows the world production of bauxite and aluminium between 1966 and 1978.

7.3 Historical development

As early as the 1700s it was realised that an unknown metal was present in certain clay soils. A very rich source was found in Southern France near the town of Les Baux, from which bauxite gets its name. Unsuccessful attempts to extract the metal from the clay were made by several European chemists, including Sir Humphrey Davy who in 1808 gave it the name aluminium.

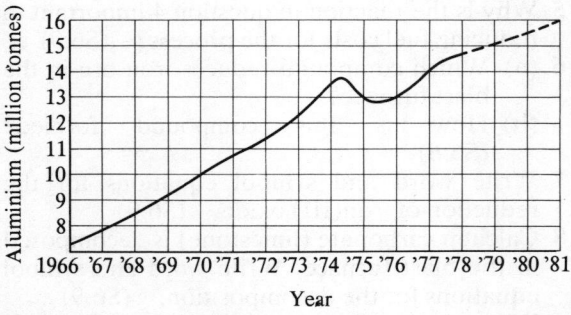

Figure 7.1. World production of aluminium between 1966 and 1978. Estimated after 1978 (broken line).

Table 7.1 World production of bauxite and alumina between 1966 and 1978.

Year	Bauxite (thousand tonnes)	Alumina (thousand tonnes)
966	41057.4	14784.4
1967	45369.2	16442.3
1968	47230.4	17472.9
1969	55457.9	19746.1
1970	60612.3	21196.4
1971	66796.6	22775.7
1972	70805.9	23611.5
1973	74866.2	26824.4
1974	83935.1	28734.0
1975	77045.2	26442.8
1976	80491.9	27536.4
1977	84780.3	25186.0
1978	84147.2	25146.0

In 1825 Danish scientist Hans Christian Oersted finally isolated the metal. He treated alumina with carbon and chlorine, then an amalgam of potassium and mercury. This resulted in a mixture of aluminium and mercury. By boiling away the volatile (low boiling point) mercury, he was left with a tiny pellet of aluminium.

Scientists next set about trying to produce the metal in larger amounts. Napoleon III, interested in the possibilities of the metal for military purposes, supported the research of Deville in the 1850s. He was able to produce more aluminium but at £20 a kilogramme it was still considered a semi-precious metal. However in 1886, by a strange coincidence, Charles Martin Hall working in a woodshed in Ohio, USA., and Paul Héroult

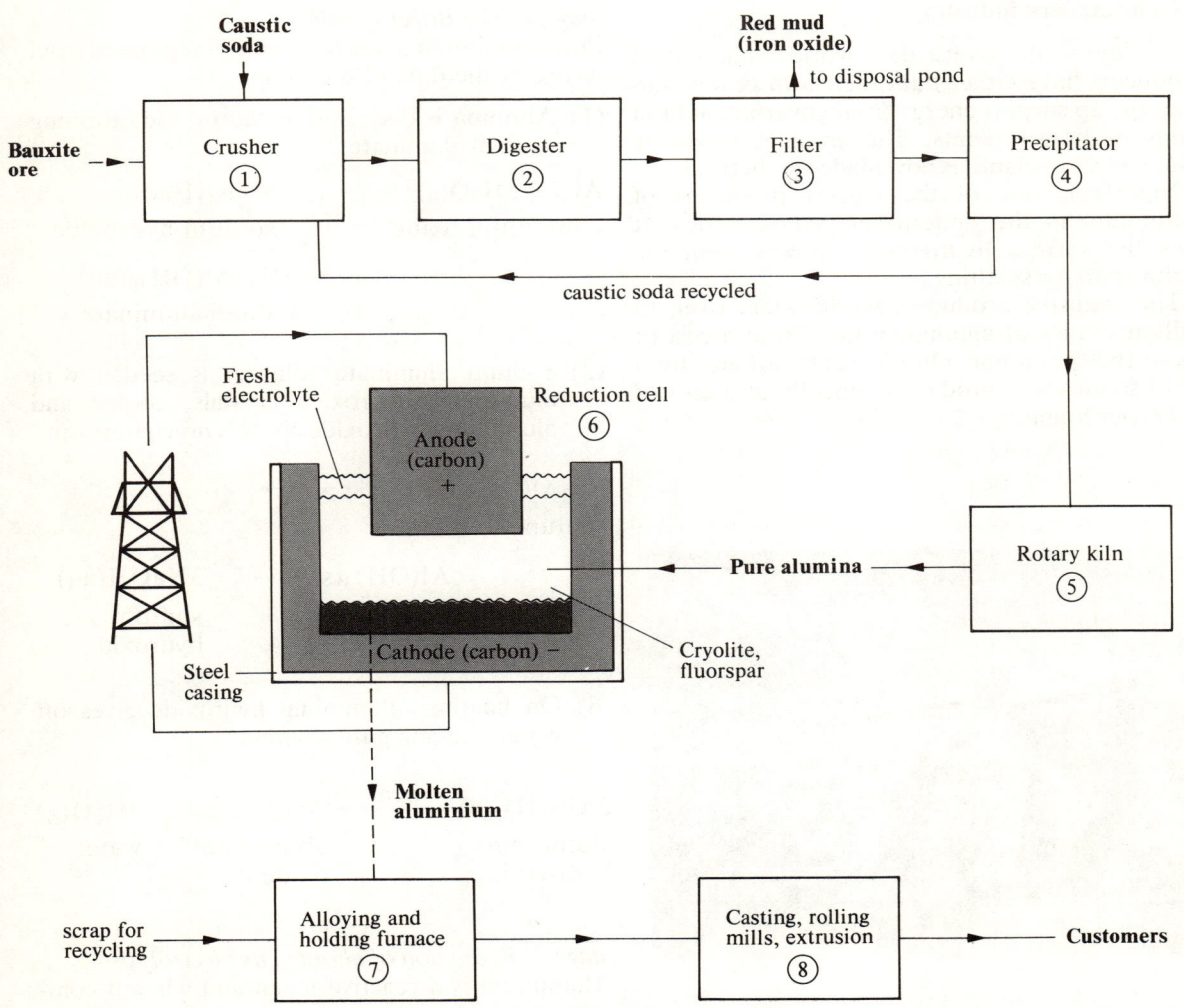

Figure 7.2. The route in a large integrated plant. Numbers refer to the stages mentioned in the text.

working in a tannery in Gentilly, France, each independently discovered a method for extracting aluminium from alumina by electrolysis. This involved passing an electric current through the alumina dissolved in a molten solvent known as **cryolite**. Both men were 22 years old when they made the discovery and both died in 1914!

The industry then took off in the USA. It needed only two main ingredients, bauxite which could be mined in Arkansas, and large quantities of electricity. When the first large scale hydro-electric plant was opened at Niagara Falls in 1895 it was not surprising then that Charles Hall's developing company was one of its first users. Other hydroelectric plants were developed when Niagara could not meet the demand.

7.4 The modern industry

Nowadays, all over the world, aluminium producers have either built their own power supplies or tap surplus energy from government built dams or power plants. The cryolite, originally mined in Greenland, is now made synthetically.

Nigeria is one of the biggest producers of aluminium in the underdeveloped countries. It uses the surplus hydroelectric power from the Volta Dam for smelting.

The industry produced, world wide, over 13 million tonnes of aluminium in 1976 at a cost of about £600 per tonne. Only 90 years earlier, a total of 15 tonnes was produced annually at a cost of £9000 per tonne.

The Volta Dam in Nigeria supplies hydro-electric power for the smelting of aluminium.

Today, the production of aluminium is carried on by a very large, highly structured industry which can be divided into 3 main sections:

1 Mining bauxite and processing to give pure alumina
2 Smelting alumina to extract aluminium
3 Treatment of metal, production of alloys, manufacture and distribution of products

7.5 Aluminium from bauxite – the chemistry of the process

The extraction process can be divided into two main stages:

Stage 1: The Bayer Process
Pure aluminium oxide (alumina) is separated from its ore by the Bayer Process.

(1) Alumina is dissolved in caustic soda forming sodium aluminate:

$Al_2O_3.3H_2O(s)$ + $6NaOH(aq)$
aluminium oxide   sodium hydroxide

$\longrightarrow 2Na_3Al(OH)_6(aq)$
sodium aluminate

(2) Sodium aluminate solution is seeded with aluminium hydroxide crystals, cooled and aluminium hydroxide crystals precipitate out:

$Na_3Al(OH)_6(aq) \longrightarrow$
sodium aluminate

$Al(OH)_3(s)$ + $3NaOH(aq)$
aluminium   sodium
hydroxide   hydroxide

(3) On heating, aluminium hydroxide gives off water, leaving pure alumina:

$2Al(OH)_3(s) \xrightarrow{\text{heat}} Al_2O_3(s)$ + $3H_2O(g)$
aluminium   aluminium   water
hydroxide   oxide

Stage 2: Reduction of alumina by electrolysis
Aluminium is a reactive metal and when it combines with other elements to form compounds the bonds between them are very strong. It requires a

42

Rows of aluminium electrolysis cells at Invergordon in Scotland. Overhead crane carries the tapping crucible which collects the molten aluminium from the electrolysis cells.

lot of energy therefore to break these bonds. Unlike iron (which is not as reactive as aluminium), aluminium cannot be extracted from its oxide by heating it with a reducing agent (see Chapter 6). Alumina can only be **decomposed** (split up) by powerful currents of electricity. The process is known as **electrolysis** (see also Chapter 8).

Alumina (Al_2O_3) is made up of aluminium ions (Al^{3+}) and oxide ions (O^{2-}). In the solid state it will not conduct electricity, but when molten it will. It is possible to get alumina in a molten state well below its melting point by dissolving it in a mixture of molten cryolite (Na_3AlF_6) and fluorspar (CaF_2). If a large electric current is passed through this mixture via carbon electrodes dipping into it, the alumina is decomposed into its ions. The Al^{3+} ions are discharged at the negative electrode (cathode) giving aluminium metal and

$$Al^{3+} \quad + \quad 3e^- \longrightarrow Al(l) \qquad \text{anode}$$

$$2O^{2-} \longrightarrow O_2(g) \quad + \quad 4e^- \qquad \text{cathode}$$

the oxide ions are discharged at the positive electrode (anode) where they combine to give oxygen.

7.6 The manufacturing process

It is very common nowadays for the bauxite ore to be processed in the country where it is mined and then shipped as pure alumina to the smelting plants. Figure 7.2 shows the stages in an integrated plant.

Stage 1: Purifying bauxite – Optional Study

Step 1: Crushing and mixing. Bauxite ore is crushed and mixed with caustic soda and pumped into digesters.

Step 2: Digestion. Caustic soda dissolves alumina under high pressure and heat to form sodium aluminate

$$Al_2O_3.3H_2O(s) \quad + \quad 6NaOH(aq)$$
$$\longrightarrow 2Na_3Al(OH)_6(aq)$$

Step 3: Settling and filtering. Sodium aluminate remains in solution, and impurities, mainly red iron(III) oxide, are allowed to settle and then pumped as red mud to disposal ponds. The solution is then filtered to further purify it.

Step 4: Precipitation and crystallisation. The sodium aluminate is pumped to a huge precipitating tank where it is further cooled, stirred and **seeded** with crystals of aluminium hydroxide. The sodium aluminate decomposes to form pure aluminium hydroxide which forms larger crystals by growing on the seed crystals.

$$Na_3Al(OH)_6(aq) \xrightarrow{\text{seed}}$$
$$Al(OH)_3(s) \quad + \quad 3NaOH(aq)$$

The sodium hydroxide is recycled to Step 1.

Step 5: Drying. The crystals of sodium hydroxide are then filtered off, washed and dried by roasting in a rotary kiln at about 1000°C. Water is driven off leaving pure alumina.

$$2Al(OH)_3(s) \xrightarrow{1000°C}$$
$$Al_2O_3(s) \quad + \quad 3H_2O(g)$$

Stage 2: Extraction of aluminium and processing the metal

Step 6: Reduction of alumina by electrolysis. The reduction cell or pot is lined with a layer of carbon which acts as the cathode of the cell. The anode is also carbon. The current is supplied through the anode via aluminium busbars. Each pot can measure up to $9 m \times 4 m \times 0.5 m$ and may take up to 22 anodes. A modern smelter may operate over 350 pots with a pot line current of 142 000 amperes. The cell is filled with molten electrolyte (cryolite + fluorspar) to which is added the alumina from Step 5. The alumina dissolves in the molten electrolyte and on the passage of the current electrolysis takes place:

$$2Al_2O_3(l) \longrightarrow 4Al(l) + 3O_2(g)$$

aluminium oxide   aluminium   oxygen

The cryolite remains unchanged and fresh alumina is added as required. The oxygen given off attacks the carbon anode and burns it away to carbon monoxide and dioxide. An anode lasts for about 3 weeks before it needs to be replaced.

Step 7: Blending and alloying. Aluminium is continuously deposited on the bottom of the cell and is siphoned out at regular intervals into crucibles. It is then transferred to holding furnaces where blending and alloying of the metal is carried out, e.g. high strength alloys for air, marine and road transport and structural engineering are produced such as:

Duralumin	Al 95%, Cu 4%, Mn, Mg, Fe, Si, 1%
Magnalium	Al 70%, Mg 30%

Step 8: Casting, rolling, extrusion processes. Molten aluminium is cast into various types of ingots – for rolling into sheet metal and **extrusion** processes (see Figure 7.3). Table 7.2 shows the quantities of raw materials that are needed in order to produce one tonne of the metal.

44

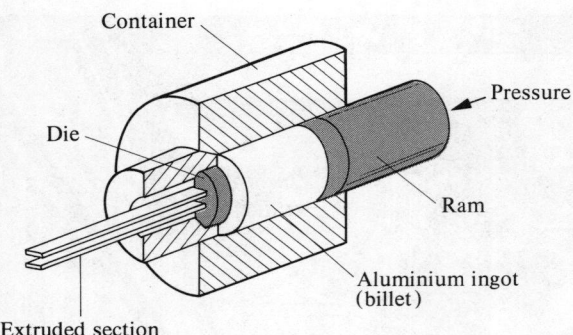

Figure 7.3. The extrusion process in which ingots of aluminium are forced through dies to produce sections of varying shapes.

Table 7.2   The various quantities of raw materials and electricity needed to produce 1 tonne of aluminium.

Material/Energy	Quantity
Bauxite	5 tonnes
Carbon anodes	0.6 tonnes
Fuel oil	0.45 tonnes
Caustic soda	0.08 tonnes
Cryolite/fluorspar	0.05 tonnes
Electricity	17000 kWh

7.7 Properties and uses of aluminium

Some of the many properties and uses of aluminium and its alloys are shown in Table 7.3.

Table 7.3 Some of the many properties and corresponding uses of the versatile metal aluminium and its alloys.

Properties	Uses
Forms light (low density), high strength alloys	Aircraft, spacecraft, ships, trains, cars, engine parts, window frames, roofing, structural engineering, construction, scaffolding, ladders
Non-toxic	Food making equipment, brewing equipment, food packaging, kitchen, foil, chemical plants
Good conductor of heat and electricity	Cooking utensils, electric kettles, overhead electricity cables
Good reflector of heat and light	Silvering large telescopes, mirrors, paints
Good reducing agent	Reducing agent, e.g. reduction of chromium oxide, steel manufacture
Non-magnetic	Navigational equipment

7.8 Environmental and social considerations

Destruction of the landscape

Open pit mining of bauxite gives rise to the same problems discussed in Chapter 6 with the mining of iron ore. Huge red basins, valleys and ridges in such countries as Australia and Jamaica bear witness to the world's thirst for aluminium. Land reclamation programmes are in progress.

Red mud ponds are another problem of the aluminium industry in regions where bauxite is purified by the Bayer Process. The main impurity is red iron(III) oxide which forms a red mud and is pumped to disposal ponds. These ponds take up vast areas of land and are very unsightly. Dykes must be built to hold the mud and there is always a danger of dyke failure. Disposal of the mud into the sea could have serious effects on aquatic life.

Pollution from the smelting process

Fluoride emissions – Some of the exhaust gas from the reduction cells is scrubbed with water and the effluent, containing fluoride (from the electrolyte), is discharged into waterways. The remaining exhaust gases are dry cleaned and discharged through tall chimneys into the atmosphere. Fluoride emissions have been known to blight vegetation and cause cattle to go lame. Research is being carried out to investigate the use of chloride in place of fluoride.

At smelters which obtain electrical power from coal burning power stations, sulphur dioxide emissions and large steam clouds are discharged through tall chimneys. Fly ash and furnace ash also produced in coal burning is mixed with sea water and pumped to settling lagoons.

How an aluminium smelter helped to revitalise a depressed region – Optional Study
A little over 10 years ago Britain had no large scale aluminium smelting industry. It imported over 90% of its aluminium. In 1968 Alcan Aluminium (UK) Ltd. built a smelter at Lynemouth in the North-east of England and went into production in 1972.

The choice of Lynemouth for the smelter. Refer to Chapter 1 for a general discussion of the factors involved in choosing a site for a chemical plant then look at Figure 7.4.

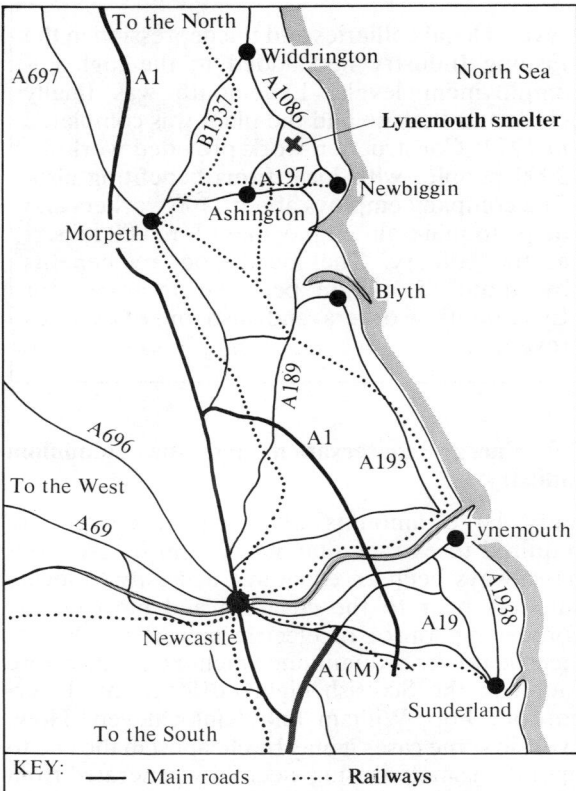

Figure 7.4. Map showing the region in the North-east of England where the Alcan Lynemouth Smelter is located. Note the very good transport network – road, rail and sea.

An **aluminium smelter** requires:

(1) a large supply of electric power: in the absence of hydroelectric power, then a ready supply of the following is needed:
    (a) coal, oil or gas.
    (b) cooling water.
(2) raw materials for the smelter, i.e. alumina, petroleum coke, cryolite and fluorspar.
(3) a good transport network which is needed to bring in raw materials to the site and take out products by road, rail and sea.
(4) a workforce.
(5) financing for the project, e.g. private investment, government grants to development areas.

At the time when the decision was being considered for the site during 1967–68 the North-east of England was a depressed region and earmarked for development. The closure of several local collieries and the depression in the Fishing Industry had added to the high unemployment levels. Lynemouth was finally chosen as the site and the plant was completed in 1972. Construction work provided work for 2000 people, with local firms benefiting also. The company employs about 1100 workers and helps to maintain employment for 1000 miners at the colliery. The local economy benefits by about £5 million per year in wages for Lynemouth workers and also provides rates revenue.

on Third World countries for supplies of bauxite just as they have done for oil.

We have already mentioned that we are likely to see more aluminium used in vehicles in order to conserve energy. Just as in the past clumsy wooden machines were replaced with iron and steel versions, we are replacing our heavy steel machines with lighter aluminium ones. Man has divided his early history into the Stone, Bronze and Iron Ages. The nineteenth century may be thought of as the Coal or Steel Age. It is quite possible that our age will be looked on as the Aluminium Age.

7.9 Energy conservation and the aluminium industry

Very large amounts of electrical energy are required to extract aluminium from its ore, so it has always been an economic necessity to locate smelters near to the cheapest and most energy conserving source of electric power, i.e. hydroelectric power. Aluminium smelters are therefore found in the Scottish highland locations: Invergordon, Fort William and Kinlochleven. However, in some cases it may be cheapest in the end to operate a smelter using electricity generated from coal, oil or gas. This is because even though the power may cost more to produce, the smelters can be located in populated areas near to the markets for the products. This reduces transport costs and the cost of building roads and houses.

However, once produced, aluminium can be recycled saving up to 95% of the original cost of producing it. In the USA especially, the recycling of drink cans is now big business, earning Americans over £20 million in 1978. It also helped reduce the litter problem.

By using more aluminium in motor vehicles their weight is reduced and hence the petrol consumption falls. It is estimated that by reducing a car's weight by 200 kg it could save on average 60 gallons of petrol each year. (See reference in Chapter 2 to the British Leyland ECV-2 – a car with an aluminium and plastic body.)

7.10 The future

Research is being undertaken to develop methods for extracting aluminium from lower grade bauxite, kaolin (clay), oil shale, and the fly ash from coal burning furnaces. However, these methods will require large amounts of energy. At the present time Western countries depend mostly

7.11 Questions on the text sections

1 Why does aluminium not corrode away like iron does? (S7.1)
2 What percentage of bauxite is pure alumina? (S7.2)
3 In which countries are the world's major deposits of aluminium ore? (S7.2).
4 What technique is used for getting the ore from the ground? (S7.2)
5 Why do you think Napoleon III encouraged research into extracting the metal? (S7.3)
6 Name the two men who first isolated aluminimum by electrolysis. (S7.3)
7 Why was aluminium smelting carried on at Niagara Falls? (S7.3)
8 What missing words or symbols are represented by the letters A to J? (S7.5)

Aluminium is extracted from its ore known as (A). . . . The ore must first be purified by the (B). . . . Process to form alumina which has the formula (C). . . . The alumina is dissolved in molten (D). . . . and (E). . . . and electrolysed in a cell which has electrodes made of (F). . . . Aluminium is deposited at the (G). . . . and oxygen at the (H). . . . Some of the oxygen produced reacts with the material of the electrode to form the gases (I). . . . and (J). . . .

9 Write out the word and symbol equations for the reaction of caustic soda with alumina during the purification of the ore. (S7.5)
10 Write out the ionic equations for the reactions occurring at the anode and cathode during electrolysis of aluminia. (S7.5)
11 For what reason are the subtances D and E in Question 8 added to the alumina during electrolysis? (S7.5)

12 Calculate the percentage of aluminium in pure alumina. (Atomic masses: $Al = 26$, $O = 16$.) (S7.2)

13 Name one alloy of aluminium used in the Aircraft Industry and state which other elements it contains. Give two properties of the alloy which make it useful for this purpose. (S7.7)

7.12 Questions on the figures and tables

1 In which year was the largest quantity of bauxite used? (T7.1)

2 Comment on the trends in bauxite and alumina production between 1966 and 1978. (T7.1)

3 What was the world annual production of aluminium metal in the years 1967 and 1974? (F.7.1).

4 What percentage of the total world production of aluminium did the UK contribute in 1978 if it produced 0.5 million tonnes? (F.7.1)

5 What percentage of the bauxite mined in 1970 was converted to aluminium? (S7.2 and T7.1)

6 What is the total quantity of raw materials needed to produce 1 tonne of Al? (T7.2)

7 Make a list of the uses of aluminium and state on which properties each use depends. (T7.3)

7.13 Longer questions

1 Refer to Table 7.1 again and plot graphs to show the world annual production of bauxite and alumina during 1966–78. Ignoring the hump in the curve in 1974, project your graph to estimate the production of these materials between 1978 and 2000. Use this graph to predict the tonnage of aluminium likely to be required in 1990.

2 Red mud ponds are a great problem in Arvida, near Quebec City, Canada, which is the site of a huge Alcan Company bauxite refining plant and smelter. Do you think that the bauxite refining process should be transferred to sites next to the bauxite mines which are generally in isolated areas in tropical countries? Give reasons for your answer. What objections do you think that the Alcan Company would have to this suggestion?

3 Eleven countries (many relatively poor) have formed the International Bauxite Association, controlling over 75% of all bauxite production. Is this a good thing for the countries in the Association? What effect is it likely to have on the future of aluminium? Bearing in mind the increase in oil prices in recent years and the hold that OPEC (a similar organisation producing oil) has on Western industrial societies, what steps are bauxite buying nations likely to take in the future to avoid similar problems?

4 Suppose the government insisted on the recycling of aluminium cans, bottle tops etc. What reasons do you think they could give to convince the public to co-operate?

5 In what ways could the use of aluminium conserve energy?

6 Read the optional study and working on your own or in small groups consider the following problem. Imagine you are part of a team of chemical engineers who have finally decided that Lynemouth should be the site for your company's new aluminium smelter. Prepare a brief report for the company directors giving all your reasons for the choice of this site.

7 Copy Figure 7.2 – Route in a Large Integrated Plant.
Explain briefly what happens at each stage of the process.

8 Chemicals from salt

8.1 Introduction

Sodium chloride (salt) is found in several areas of Britain, but the only areas which supply the chemical to industry are shown in Figure 8.1, together with the sites of the factories. Only from the Winsford mine is solid sodium chloride mined; in other areas it is extracted by dissolving it in water and pumping up the resulting brine.

Apart from sodium chloride itself, salt is converted into many important elements and compounds such as chlorine, sodium hydroxide, sodium carbonate, hydrogen, sodium, hydrogen chloride, hydrochloric acid, sodium hydrogen carbonate and sodium chlorate. Of these, the first three are probably the most important. The important industrial processes are shown below.

(1) Diaphragm cell (Chlor–Alkali Process)
chlorine
Mercury cell sodium hydroxide
hydrogen
(2) Solvay Process sodium carbonate
(3) Downs cell sodium and chlorine

Other important chemicals can be made from the products of these processes.

8.2 The Chlor-Alkali Process

The process was begun by the Castner-Kellner Company in Runcorn in 1897. At first the important product was sodium hydroxide, used for the manufacture of soap, but a glance at Figure 8.2 shows how the importance of chlorine has increased dramatically since 1900. Figures 8.3 and 8.4 show what these two chemicals are used for in the UK now.

As vinyl chloride is used to make PVC, chloromethanes are used as 'anti-knock' additives for petrol, and chlorine based solvents are non-flammable, the importance of chlorine can be seen.

8.3 Chemistry of the process

When sodium chloride is dissolved in water it splits up into **ions** (charged particles). Four ions are

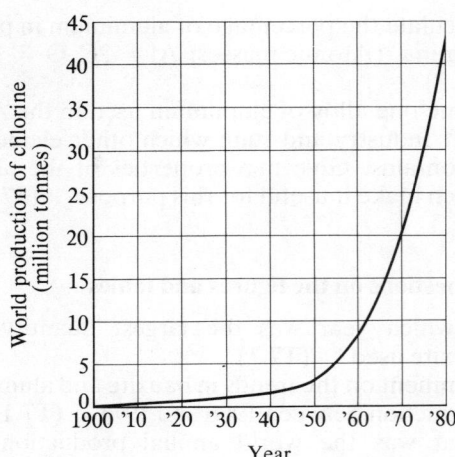

Figure 8.2. Graph showing how the demand for chlorine has grown over the past few decades, indicating the importance of the element to modern society.

present, sodium ions and chloride ions from sodium chloride and hydrogen ions and hydroxyl ions from water. This solution can conduct electricity and is called an **electrolyte.**

If metal plates (**electrodes**) connected to a source of electricity are put into the solution, the positively charged ions (**cations**) move towards the negatively charged electrode (**cathode**). The negatively charged ions (**anions**) move towards the positively charged electrode (**anode**). This is called **electrolysis**. Only one ion can be discharged at each electrode.

$$NaCl \longrightarrow Na^+ + Cl^-$$
$$H_2O \longrightarrow H^+ + OH^-$$

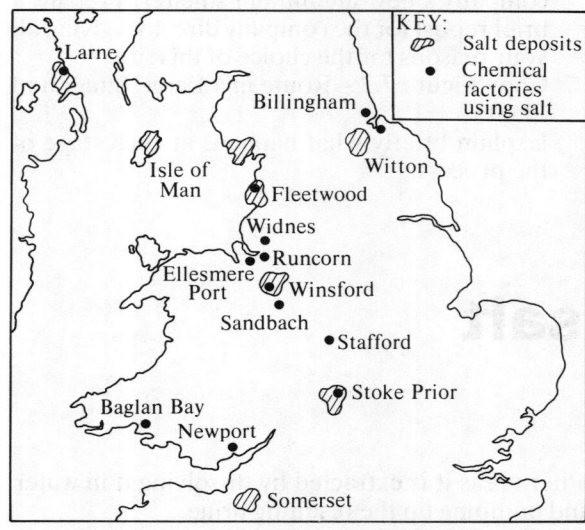

Figure 8.1. The salt deposits which are currently being worked in Great Britain and the factories which use salt (sodium chloride) as a raw material.

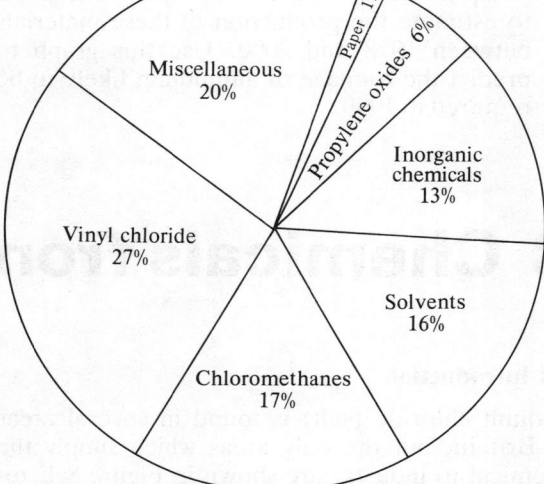

Figure 8.3. The uses of chlorine. Note how much of the element is combined with organic chemicals.

48

Discharge of ions – Optional Study
Which ions are discharged normally depends on three criteria.

The ions present. The ease of discharge is shown here with the most easily discharged at the right-hand side:

$K^+, Ca^{2+}, Na^+, Mg^{2+}, Al^{3+}, Zn^{2+}, Fe^{2+}, Pb^{2+}, H^+, Cu^{2+}, Hg^{2+}, Ag^+$

→ more easily discharged

$SO_4^{2-}, NO_3^-, Cl^-, Br^-, I^-, OH^-$

So in the electrolysis of a solution of a copper salt (Cu^{2+} and H^+) copper is discharged at the cathode, but in the electrolysis of a sodium salt solution (Na^+ and H^+), hydrogen is discharged.

Concentration. In a concentrated solution, the more concentrated ion may be discharged. In dilute sodium chloride solution, oxygen (from hydroxyl ions) will be produced at the anode, but with concentrated sodium chloride, chlorine is produced.

The electrode. A steel cathode releases hydrogen from sodium chloride solution, but a mercury cathode releases sodium.

At the cathode, hydrogen ions pick up electrons to form hydrogen gas.

$$2H^+(aq) + 2e^- \longrightarrow H_2(g)$$

At the anode, chloride ions give up electrons to form chlorine gas.

$$2Cl^-(aq) \longrightarrow Cl_2(g) + 2e^-$$

As hydrogen ions and chloride ions are being removed, the two ions left are sodium (Na^+) and hydroxyl (OH^-), so sodium hydroxide solution ($NaOH(aq)$) is formed.

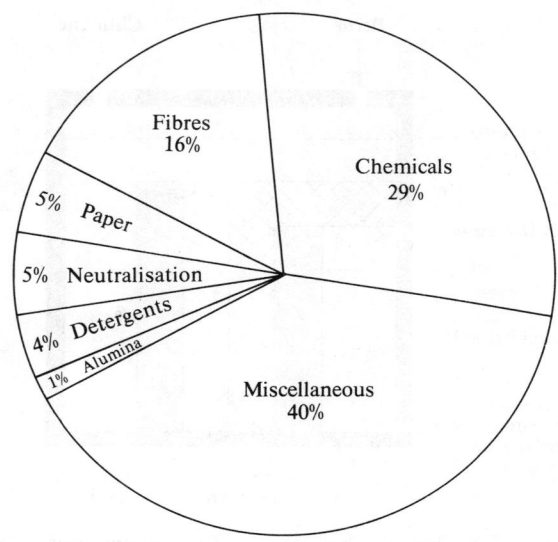

Figure 8.4. Uses of sodium hydroxide. This used to be the most important product from salt, and still is in the Third World, but has now been overtaken by chlorine.

8.4 The modern manufacturing process

Two types of cell are used, both of which have good and bad points as shown in Table 8.1.

In both cells the idea is to separate the chlorine from the hydrogen (with which it forms an explosive mixture) and the sodium hydroxide (which forms sodium hypochlorite).

8.5 The Diaphragm cell

Figure 8.5 shows a simplified Diaphragm cell.

The **brine** contains 25% by weight of sodium

Table 8.1 A comparison of the Diaphragm and Mercury cells. The Mercury cell is more expensive to set up, but the products are much purer.

Cell	Construction	Operation	Sodium hydroxide solution
Diaphragm	Simple, inexpensive	Diaphragm replaced regularly. Needs constant current. Voltage 3.8 V	Concentration weakens. Contains salt
Mercury	Expensive to construct. Mercury expensive	Mercury toxic. Operates over wide current load. Voltage 4.5 V	Stays at constant concentration. Pure

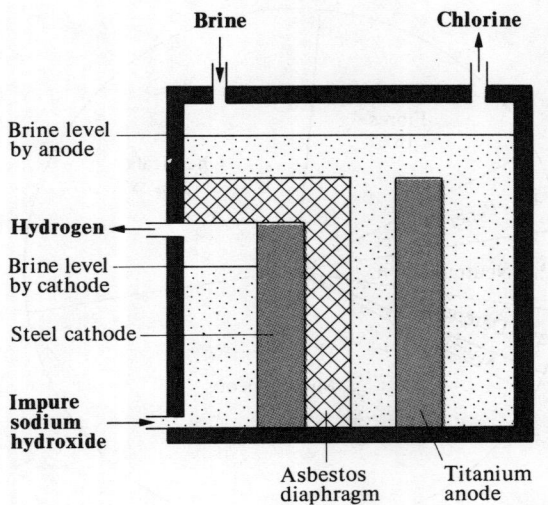

Figure 8.5. The diaphragm cell. The sodium hydroxide produced by this method contains salt and has to be purified.

chloride (75% water). Half of the chloride ions are converted to chorine gas at the titanium anode.

$$2Cl^-(aq) \longrightarrow Cl_2(g) + 2e^-$$

As the brine in the anode compartment is higher than that in the cathode compartment, some seeps through the asbestos diaphragm to replace the liquid run off.

At the steel cathode the hydrogen ions are converted into hydrogen gas.

$$2H^+(aq) + 2e^- \longrightarrow H_2(g)$$

The brine electrolyte is converted to 'cell liquor', containing 10% sodium hydroxide and 15% sodium chloride. Evaporation crystallizes out the sodium chloride, leaving a solution containing 50% sodium hydroxide and under 1% sodium chloride.

8.6 The Mercury cell

The Mercury cell also has a titanium anode, but a mercury cathode. At the anode sodium ions are converted to sodium.

$$Na^+(aq) + e^- \longrightarrow Na(s)$$

The sodium forms an **amalgam** (solution) in the mercury and does not react with the solution. The

liquid mercury is run off (the floor of the cell slopes) and the amalgam is allowed to react with distilled water over activated graphite blocks. It produces sodium hydroxide solution and hydrogen gas.

$$2Na/Hg(l) + 2H_2O(l) \longrightarrow$$

sodium   water
amalgam

$$\qquad 2NaOH(aq) + H_2(g) + 2Hg(l)$$

     sodium   hydrogen   mercury
     hydroxide

The mercury is returned to the cell (see Figure 8.6). The reaction at the anode is the same as in the Diaphragm cell.

$$2Cl^-(aq) \longrightarrow Cl_2(g) + 2e^-$$

The advantage of this cell is that the products are all pure (apart from containing water).

8.7 Uses of products

The products are in the ratio 2 NaOH : 1 H_2 : 1 Cl_2, which means that more sodium hydroxide is made than is required in this country. The Third World, however, does require more sodium hydroxide than chlorine and some is converted to sodium carbonate.

The hydrogen is mainly used for reduction of organic compounds and the hydrogenation of fats.

Chlorine is very important, as can be seen from its uses listed in Table 8.2.

8.8 The Ammonia-Soda Process

Sodium carbonate was originally made by the Leblanc Process, which had three main disadvantages. These were:

(1) the large energy requirement.
(2) the multi-stage process using small batches which needed a lot of labour.
(3) the large amount of pollution.

The problem was finally overcome with the invention of the Solvay Process in 1865.

The sodium carbonate is made in two forms: heavy ash, which is denser and easier to transport, and light ash, which is cheaper and has no calcium ions. Their uses are shown in Table 8.3.

Table 8.2. The uses of chlorine, showing the importance of compounds containing the element.

Reaction	Product	Uses
	Chlorine	Sterilisation
Cl_2 + organic chemicals	Organic chlorides	Weedkiller, insecticide
$CH_2 = CH_2 \xrightarrow{Cl_2} CH_2(Cl)CH_2Cl \xrightarrow{-HCl} CH_2 = CHCl$	Polyvinylchloride	Plastics
Cl_2 + alkanes	Chloroalkanes	Solvents
$Cl_2(aq) + CH_3CH = CH_2 \xrightarrow{-HCl} CH_3CH\overset{O}{-\!\!\!\triangle\!\!\!-}CH_2$	Propylene oxide	Drugs, plastics, brake fluid
$H_2 + Cl_2 \rightarrow 2HCl$	Hydrochloric acid	Chemicals
$2NaBr + Cl_2 \rightarrow 2NaCl + Br_2$	Bromine	Chemicals
$NaOH + Cl_2 \rightarrow NaOCl + HCl$	Sodium hypochlorite	Bleach
Cl_2 + inorganic chemicals	Inorganic chlorides	Chemicals

Table 8.3 The uses of heavy and light ash, the two main forms of sodium carbonate produced.

Heavy ash		Light ash	
Glass containers	63%	Heavy chemicals	57%
Other glass	20%	Oil and fats	16%
Sodium silicate	9%	Textiles	3%
Miscellaneous	8%	Dyes and colours	3%
		Food and drink	3%
		Fine chemicals	3%
		Miscellaneous	15%

8.9 The chemistry of the process

The reaction

$$2NaCl(aq) \quad + \quad CaCO_3(s)$$
$$\text{sodium} \qquad\quad \text{calcium}$$
$$\text{chloride} \qquad\quad \text{carbonate}$$

$$\rightleftharpoons \quad CaCl_2(aq) \quad + \quad Na_2CO_3(aq)$$
$$\qquad\quad \text{calcium} \qquad\quad \text{sodium}$$
$$\qquad\quad \text{chloride} \qquad\quad \text{carbonate}$$

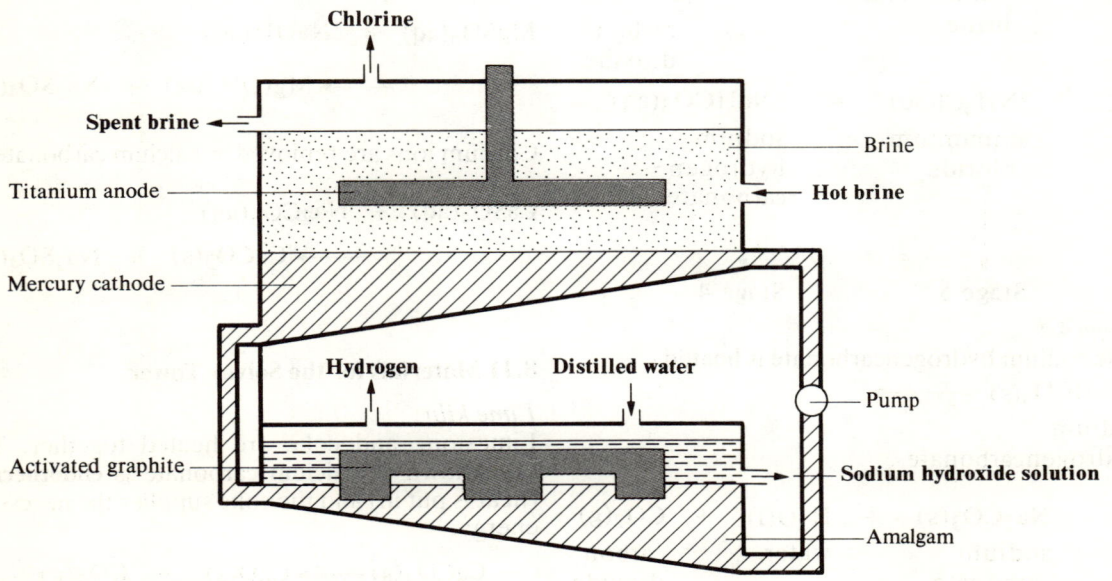

Figure 8.6. The mercury cell. The mercury is pumped round in a circle, picking up the sodium as an amalgam, then releasing it into the distilled water as pure sodium hydroxide solution.

51

is far more likely to go from right to left than from left to right (the calcium carbonate is insoluble, so is removed from the reaction), so it cannot be used directly. The actual process is in five stages.

Stage 1
Calcium carbonate is heated in a lime kiln.

$$CaCO_3(s) \longrightarrow CaO(s) + CO_2(g)$$

calcium carbonate   calcium oxide   carbon dioxide

         ↓ Stage 2   ↓ Stage 3

Stage 2
The calcium oxide is dissolved in water.

$$CaO(s) + H_2O(l) \longrightarrow Ca(OH)_2(aq)$$

calcium oxide   water   calcium hydroxide

         ↓ Stage 5

Stage 3
The carbon dioxide is reacted with ammoniacal brine.

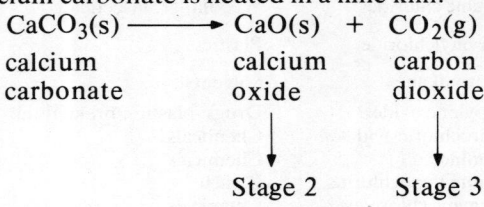

$$2NaCl(aq) + 2NH_3(aq) + 2H_2O(l)$$

sodium chloride   ammonia   water

ammoniacal brine $+ 2CO_2(g)$ carbon dioxide

$$\rightleftharpoons 2NH_4Cl(aq) + 2NaHCO_3(aq)$$

ammonium chloride   sodium hydrogen-carbonate

    ↓ Stage 5   ↓ Stage 4

Stage 4
The sodium hydrogencarbonate is heated.

$$2NaHCO_3(s) \longrightarrow$$

sodium hydrogencarbonate

$$Na_2CO_3(s) + H_2O(l) + CO_2(g)$$

sodium carbonate   water   carbon dioxide

         ↓ Stage 3

Stage 5
The ammonium chloride and calcium hydroxide are reacted.

$$Ca(OH)_2(aq) + 2NH_4Cl(aq)$$

calcium hydroxide   ammonium chloride

$$\rightleftharpoons CaCl_2(aq) + 2NH_3(g) + 2H_2O(l)$$

calcium chloride   ammonia   water

         ↓ Stage 3

As the amount of carbon dioxide and ammonia remains constant, these can be recycled. Sodium carbonate is produced at Stage 4 and the other products are shown in Figure 8.7.

8.10 Sources for the modern manufacturing process
Brine is obtained by pumping water into salt deposits and bringing the solution (containing magnesium and calcium sulphate impurities) to the surface. Magnesium ions are removed as magnesium hydroxide.

$$MgSO_4(aq) + 2NaOH(aq)$$
$$\longrightarrow Mg(OH)_2(s) + Na_2SO_4(aq)$$

Calcium ions are removed as calcium carbonate.

$$CaSO_4(aq) + Na_2CO_3(aq)$$
$$\longrightarrow CaCO_3(s) + Na_2SO_4(aq)$$

8.11 Materials for the Solvay Tower
Lime kiln
Limestone and coke are heated together. The breakdown of calcium carbonate is **endothermic** (heat is put in) and the coke supplies the necessary energy.

$$CaCO_3(s) \longrightarrow CaO(s) + CO_2(g)$$

The carbon dioxide is passed into the Solvay Tower.

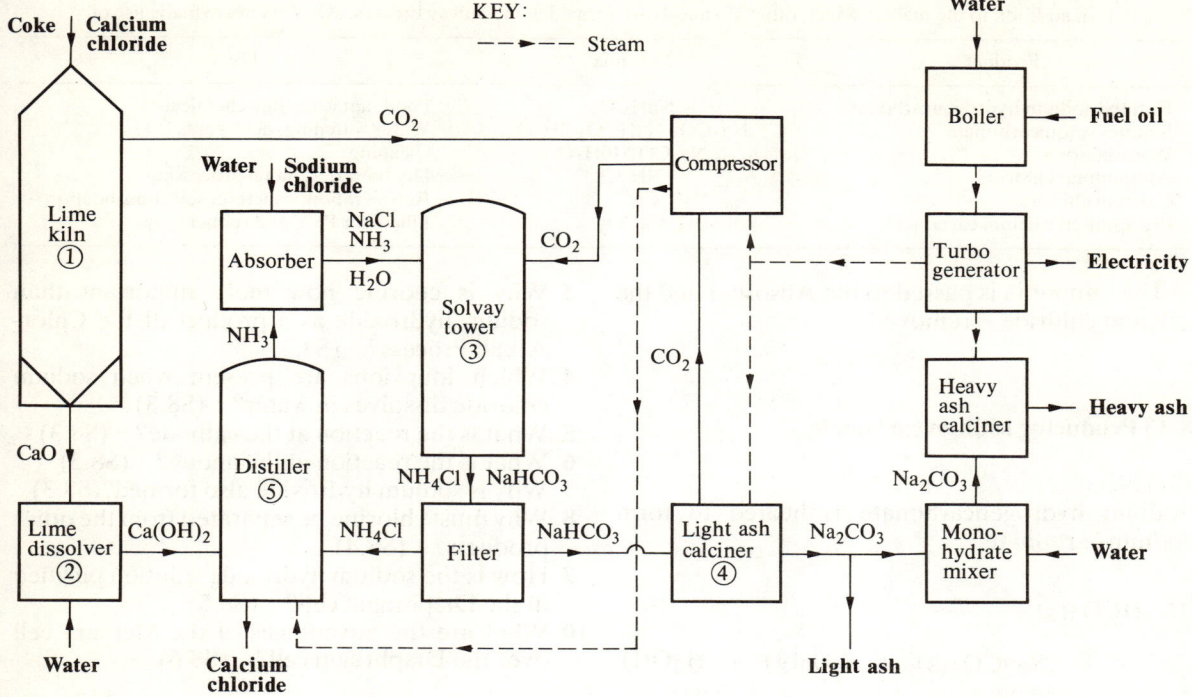

Figure 8.7. The route in the modern Solvay process. Note how many by-products are recycled to obtain maximum use of them. Numbers refer to stages mentioned in the text.

Lime dissolver

The calcium oxide is dissolved in water. As the strength of the solution is not enough, more calcium oxide is added until a concentrated suspension of calcium hydroxide in water is formed.

$$CaO(s) + H_2O(l) \longrightarrow Ca(OH)_2(aq + s)$$

The calcium hydroxide is passed into the Distiller

Absorber

Brine is mixed with ammonia from the Distiller to form ammoniacal brine.

$$NaCl(aq) + H_2O(l) + NH_3(g) \longrightarrow$$
$$NaCl(aq) + H_2O(l) + NH_3(aq)$$

The ammoniacal brine is passed into the Solvay Tower.

8.12 Producing sodium hydrogencarbonate

Solvay tower

Ammoniacal brine reacts with carbon dioxide to form sodium hydrogencarbonate and ammonium chloride.

$$NaCl(aq) + H_2O(l) + NH_3(aq) + CO_2(g)$$
$$\longrightarrow NaHCO_3(aq) + NH_4Cl(aq)$$

The liquid is cooled until the sodium hydrogencarbonate forms crystals of a carefully controlled size. These are filtered off and sent to the Light Ash Container. The ammonium chloride filtrate is sent to the Distiller.

Distiller

The ammonium chloride reacts with calcium hydroxide.

$$Ca(OH)_2(aq) + 2NH_4Cl(aq) \longrightarrow$$
$$CaCl_2(aq) + 2NH_3(g) + 2H_2O(l)$$

53

Table 8.4 In addition to the main product, other chemicals are formed in the Solvay Process. All of them are made use of.

Product	Formula	Uses
Refined sodium hydrogencarbonate	$NaHCO_3$	Food, antacids, fine chemicals
Sodium sesquicarbonate	$Na_2CO_3 \cdot NaHCO_3 \cdot 2H_2O$	Water softening, detergents
Washing soda	$Na_2CO_3 \cdot 10H_2O$	Cleaning
Ammonium chloride	NH_4Cl	Dry batteries, metal processing
Calcium chloride	$CaCl_2$	Refrigeration, concrete, soil consolidation
Precipitated calcium carbonate	$CaCO_3$	Fillers for PVC and rubber

The ammonia is passed to the Absorber and the calcium chloride is removed.

8.13 Producing sodium carbonate

Calciners
Sodium hydrogencarbonate is heated to form sodium carbonate.

$$2NaHCO_3(s) \longrightarrow$$
$$Na_2CO_3(s) + CO_2(g) + H_2O(l)$$

The carbon dioxide is passed back to the Solvay Tower. The sodium carbonate produced is light ash. To form heavy ash, water is added to form the monohydrate ($Na_2CO_3.H_2O$) which is then dehydrated to form larger, denser particles.

8.14 The Downs cell

See Figure 8.8. As no water is present, sodium metal is formed. Added calcium chloride lowers the melting point of sodium choride from 800°C to 600°C.

Cathode reaction: $Na^+ + e^- \rightarrow Na(l)$
Anode reaction: $2Cl^- \rightarrow Cl_2(g) + 2e^-$

The low density of sodium means that it floats up to be collected in an inverted trough, while chlorine is collected by a hood.

Sodium is used as a coolant in nuclear reactors, in sodium vapour street lamps, as a reducing agent and in the manufacture of sodium salts.

8.15 Questions on the text sections

Chlor-Alkali Process
1 How is salt removed from the ground? (S8.1)
2 Which industrial processes produce chemicals from salt? (S8.1)

3 Why is chorine now more important than sodium hydroxide as a product of the Chlor-Alkali Process? (S8.2)
4 Which four ions are present when sodium chloride dissolves in water? (S8.3)
5 What is the reaction at the cathode? (S8.3)
6 What is the reaction at the anode? (S8.3)
7 Why is sodium hydroxide also formed?(S8.3)
8 Why must chlorine be separated from the other products? (S8.4)
9 How is the sodium hydroxide solution purified in the Diaphragm cell? (S8.5)
10 What are the advantages of the Mercury cell over the Diaphragm cell? (S8.6)

Ammonia-Soda Process
11 What were the disadvantages of the Leblanc Process? (S8.8)
12 What are the differences between heavy ash and light ash? (S8.8)
13 What is ammoniacal brine? (S8.9)

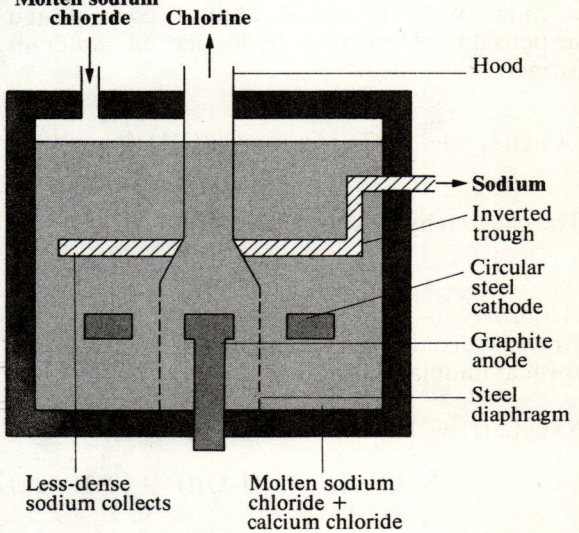

Figure 8.8. The Downs cell.

54

14 What is the reaction of ammoniacal brine with carbon dioxide? (S8.9)
15 How is sodium hydrogencarbonate converted to sodium carbonate? (S8.9)
16 What happens in the lime kiln? (S8.11)
17 How is the sodium hydrogencarbonate separated from the ammonium chloride in the Solvay Tower? (S8.12)

Downs cell
18 (a) How is the melting point of sodium chloride lowered? (b) Why is this important? (S8.14)
19 How is the sodium collected? (S8.14)
20 What is sodium used for? (S8.14)

8.16 Questions on the figures and tables

1 In which areas of Britain is salt extracted? (F8.1)
2 How much chlorine was produced in 1970? (F8.2)
3 What are the differences between the Diaphragm cell and the Mercury cell? (T8.1.)
4 Make a list of the products of chlorine and their uses. (T.8.2)

5 Which raw materials are used in the Solvay Process and into which parts of the plant are they put? (F8.7)
6 Make a list of the products of the Solvay Process and their uses. (T8.3 and T8.4)

8.17 Longer questions

1 Explain what happens during electrolysis. Use the optional study to explain why hydrogen and chlorine are produced in the Diaphragm cell.
2 Draw Figure 8.6. Explain how the cell works and how the products are removed.
3 Draw Figure 8.7. Describe, with equations, what happens at each stage of the process.
4 Why do you think that more chlorine is produced commercially by the electrolysis of sodium chloride solution than the electrolysis of molten sodium chloride?
5 Explain the importance of salt to the Chemical Industry.
6 Read the optional study on Le Chatelier's Principle (Chapter 3). Explain how the formation of solid calcium carbonate in the equation

$$2NaCl(aq) + CaCO_3(s) \rightleftharpoons CaCl_2(aq) + Na_2CO_3(aq)$$

affects its concentration and hence the position of equilibrium.

Background resources for use with case studies

Slides/Filmstrips/Tapes

Modern Industrial Chemistry
Each set has 12 slides. Available from Philip Harris Biological Ltd., Oldmixon, Weston-super-Mare, BS24 9BJ.

Aluminium	(2 sets)
Sulphuric Acid	(1 set)
Petroleum Refining	(1 set)
Ammonia	(1 set)
Nitric Acid	(1 set)
Chemicals from Salt	(3 sets)
Iron and Steel	(3 sets)

Science and Technology in Society
Each set has 24 slides/frames. Available from Audio-Visual Productions, Hocker Hill House, Chepstow, NP6 5ER.

Science and Technology in Society	(1 set)
Great Industries of Britain	(1 set)
Britain's Industrial Landscape	(1 set)
The Oil Industry	(1 set)
Alternative Energy	(1 set)
Mining and Minerals	(1 set)

Only One Earth
Filmstrip or cassette. Available from Visual Publications, 197 Kensington High Street, London, W8 6BB.

The Biosphere: material cycles
Fuels and Energy
Land and Land Pollution
Water and Water Pollution
Air Pollution

Films

The Chemical Society Film Index, Education Officer, The Chemical Society, Burlington House, London W1V 0BN.

British Petroleum Film Library, 15 Beaconsfield Road, London, NW10 2LE.

I.C.I. Film Library, 15 Beaconsfield Road, London NW10 2LE.

Viscam Ltd., The Film Library, Park Hall Road Trading Estate, London, SE21 8EL. (Supply films for British Gas, British Steel)

Central Film Library, Government Buildings, Bromyard Avenue, Acton, London, W3 7JB.

National Audio-Visual Aid Film Library (E.F.V.A.), Paxton Place, London SE27 9SR.

Charts and Booklets

Imperial Chemical Industries Ltd., Schools Liaison Officer, Thames House North, Millbank, London, SW1P 4QG.

Alcan Aluminium (UK) Ltd., Southam Road, Banbury, OX16 7SN.

Institute of Petroleum Information Service, 61 New Cavendish Street, London W1M 8AR.

National Sulphuric Acid Association Ltd., Accounts Department, Piccadilly House, 16 Jermyn Street, London, SW1Y 4NF.

Conservation Society Ltd., 12A Guildford Street, Chertsey, Surrey, KT16 9BQ.

British Steel Corporation, Information Officer, 151 Gower Street, London WC1E 6BB.

British Gas, Education Liaison Officer, Room 414, 326 High Holborn, London, WC1V 7PT.

Understanding British Industry, The Resource Centre Director, Sun Alliance House, New Inn Hall Street, Oxford, OX1 2QE.

Schools Information Centre on the Chemical Industry, Polytechnic of North London, Holloway Road, London, N7 6DB.

A very detailed and useful resource list covering books, charts, films etc. relating to the Chemical Industry is available from The Joint Matriculation Board, Manchester, M15 6EU.

Index